Equipment
1993 EDITION
BUYERS
GUIDE

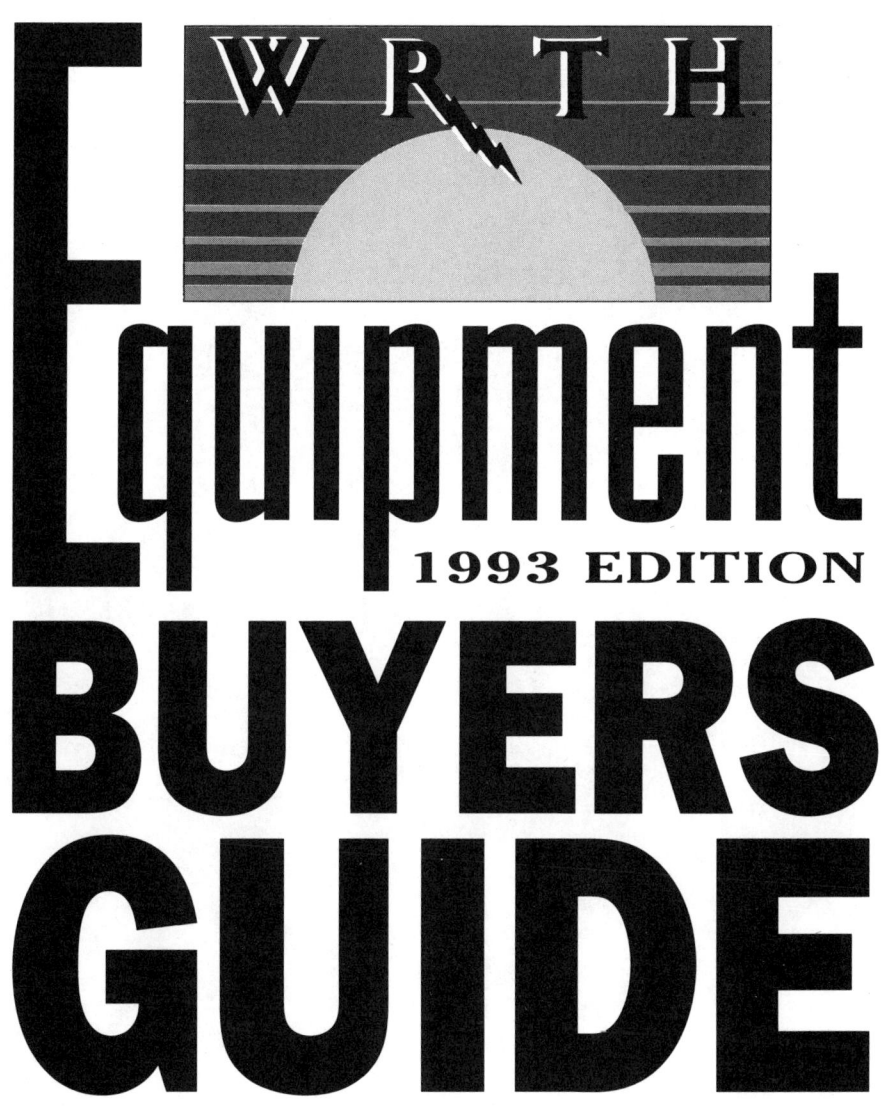

WRTH Equipment Buyers Guide
1993 Edition

BILLBOARD BOOKS
an imprint of Watson-Guptill Publications
New York/Amsterdam

Authors: Willem BOS and Jonathan Marks
Editor: Andy Sennitt
Associate Editor: Bart Kuperus
Art Director: Bob Fillie
Publisher: Glenn Heffernan

Copyright © 1993 by Billboard Books,
an imprint of Watson-Guptill Publications,
a division of BPI Communications, Inc.
1515 Broadway, New York 10036

All rights reserved. No part of this publication may be reproduced, stored in a retrieval system, or transmitted, in any form or by any means - electronic, mechanical, photocopying, recording, or otherwise - without written permission of the publisher.

Manufactured in the United States of America.

ABOUT THE AUTHORS: 26 YEARS OF TESTING

WILLEM BOS & JONATHAN MARKS

Members of the WRTH test bench team are no strangers to the world of shortwave receivers. Although the WRTH is a commercial publication, the compilation of this section is editorially independent from any pressure by advertisers. The WRTH has been testing receivers for longer than many other publications. It can now draw on the improved resources of over a quarter of a century of experience. In 1967, Willem BOS published his first equipment test in the Dutch electronics magazine "Radio Bulletin". He has worked hard to promote better understanding of consumer electronics, and has been awarded several journalism prizes as a result, such as the ELPEC prize 1988. Willem has set up one of the largest measuring laboratories in Europe dedicated to receiver and antenna measurements. Jonathan Marks began his career with Austrian Radio in 1976, joining Radio Netherlands in 1980. He created the concept of a "communications magazine" with the launch of "Media Network" in May 1981. In 1992 an edition of the programmed won a finalist certificate and Gold medal at the International Radio Festival of New York.

FOREWORD

Welcome to this brand new venture from the publishers of the World Radio TV Handbook. In the last five years there has been a lot of feedback to the Amsterdam editorial offices of the WRTH. Each year the receiver section of the World Radio TV Handbook critically reviews the new receivers that have appeared over the last 12 months. However with an ever increasing number of people discovering the world of international radio, getting an overview of what's on the market has become more difficult. By taking the reviews since 1987 and, in some cases, retesting upgraded models we've come up with this buyers guide. We've taken the opportunity to add readers' comments and include new material that hasn't been previously published. Your reaction to this new venture is requested. Remember to take part in the special contest at the back of this book.

CONTENTS:

CHAPTER ONE
WHERE TO BUY A RECEIVER 11

CHAPTER TWO
STEPS TO PICKING THE RIGHT RECEIVER 15

CHAPTER THREE
SORTING OUT RECEIVER SPECIFICATIONS. 18

CHAPTER FOUR
HARD FACTS ABOUT ACTIVE AND PASSIVE ANTENNAS 38

ANTENNAS
Alpha Delta Sloper 47
Dipole 50
Supreme Listener 50
T2FD Antenna 51
RF Systems "Magnetic Longwire Balun" 56
Dressler ARA-60 58
RF Systems DX-1 60
DX Listener 61
Sony AN-1 63
Datong AD270/AD370 65
Phase Track Liniplex Loop Antenna 66
DX-7 69
Yaesu FRA-7700 70
Interceptor Loop Antenna 71

ACCESSORIES
SP-2 Antenna Splitter 73
Palomar Amplifilter 76
Lowe PR-150 77
Palomar Channel Cleaner 78

CHAPTER FIVE
VIEW FROM THE DEVELOPING COUNTRIES 81

CHAPTER SIX
LOOPS FOR SHORTWAVE 84

CHAPTER SEVEN
FREQUENCY MANAGEMENT- THE KEY TO GOOD RECEPTION   99

CHAPTER EIGHT
RECEIVER TEST RESULTS 105

AIWA WRD 1000 187
DAK MR-101 105
DAK DMR-3000 107
Drake R8 109
Grove SW-100 114
Grundig Yacht Boy 206 115
Grundig Yacht Boy 220 116
Grundig Satellit 500 118
Grundig Satellit 700 125
Icom IC-R9000 130
Icom IC-R100 133
Icom IC-R1 138
Icom IC-R72 140
JRC NRD-535 144
Kenwood R-2000 148
Kenwood R-5000 149
Lowe SRX-50 156
Lowe HF-125 157
Lowe HF-150 163
Lowe HF-225 167
Panasonic RF-B65L 172
Panasonic RF-B45 175
Philips (Magnavox) AW-3805/Sangean ATS-800 177
Philips AE-3905 180
Radio Shack DX 380 187
Roberts R808 187
Sangean ATS 818(CS) 184
Sangean ATS 808 187
Siemens RK710 182
Siemens RK631 183
Siemens RK661 187
Siemens RK670/Sangean ATS818(CS)/Tandy DX-390 184
Sony ICF-SW1S/SW1E 189
Sony ICF-SW15 194
Sony ICF-SW20 195
Sony ICF-SW7600 197
Sony ICF-SW7601 200
Sony ICF-2001D/ICF-2010 203
Sony ICF-SW77 207

Sony ICF-SW55 212
Sony CRF-V21 216
Tandy DX-390 184
Yaesu FRG-8800 220

CHAPTER NINE
SHORTWAVE RADIO IN THE CAR 225

Kenwood RZ-1 232
Philips DC777 235

CHAPTER TEN
RADIO RELATED COMPUTER SOFTWARE 240

Shortwave Broadcast Schedules 240
Radio Listener's International PC Database 241
Public Domain Software for the Macintosh 243
DX Helper/Satellite Pro 243
SW Navigator 243
Badview 1.50i for the PC 244
Macratt for the Pakratt 232 245
Seeker-PC for the Kenwood R-5000 246
Interval Signals version 1.21 246
Interval Signals On Line 248
Short Wave Log 1.10 248
Other Radio-Related Software 250
Computer Bulletin Boards 250, 253

CHAPTER ELEVEN
CONTACT ADDRESSES FOR FURTHER INFORMATION 261

Shortwave Equipment Sources 261
Military Surplus Receivers 264
Vintage Radio Societies 265
Other Sources 266
Service Sheet Sources 267
Further Sources of Receiver Information 268

CHAPTER TWELVE
WRTH READER PRIZE QUESTIONNAIRE 269

CREDITS 270

CHAPTER 1
WHERE TO BUY A RECEIVER

Fifteen years ago buying a shortwave receiver in Europe or North America was quite difficult. If you didn't live near an amateur radio outlet you probably never saw a table-top receiver. The portables on the market only offered mediocre performance and information about shortwave international broadcasting was difficult to find. If you didn't already know something about international radio, chances were that you wouldn't find out about it. Today good portable shortwave receivers are much easier to find...many electrical appliance stores stock them. And the recession in the amateur radio market has meant that these specialized shops have started to cater more for the serious shortwave broadcast listener. Note that Sony has been successful in launching a new term into the field of international broadcasting: "worldband radio". However, we note that the term "worldband" is also a Sony trademark, so we have avoided its use in this publication. We also prefer the term "international radio listening" to "shortwave listening" because more and more stations are using other media (e.g. satellites and mediumwave (AM)) to reach their listeners.

In the course of the last ten years, Radio Netherlands, Holland's external broadcasting service, has been making an annual survey of the shortwave receiver market place. By using a network of volunteers, several thousand shops have been visited over the last decade to check on the level of advice you get from behind the counter and the prices being charged. Analysis of this worldwide data means we're able to offer you a global picture of the current market. Of course, individual shops vary depending on the manager. But here is a general rule of thumb:

AIRPORT DUTY FREE

If you're travelling around, the best "DUTY-FREE" prices are at Singapore (but see below), Hong Kong and Amsterdam Schiphol (in that order). Most other airports sell a restricted range of shortwave equipment at higher prices than local shops. The worst prices were in Brussels, Geneva, and London Heathrow. With the exception of Amsterdam Schiphol, the level of salesmanship advice is VERY POOR EVERYWHERE. Salesmen want to sell you luxury gifts / video recorders, not radio sets.

SOUTH-EAST ASIA / JAPAN

In cities like Hong Kong, Bangkok and Singapore, radios are not usually displayed in the shop window with a visible price tag. It is up to you to bargain! If you allow time for shopping, and refuse to accept the first price given, you may save hundreds of dollars. You should consider possible customs duty that you may have to pay when you arrive home. Shortwave communications receivers from Yaesu, Kenwood, Icom and Japan Radio Company are usually confined to amateur radio outlets. They are rarely seen in the photo-discount outlets.

In Singapore the airport prices are fixed, and the range of shortwave equipment is restricted to portables. Shopping in the city itself is considerably cheaper. Most tours take foreign tourists to "Lucky Plaza". Most current shortwave portables are to be found there, but again, remember to bargain. The bottom price is usually 60 - 70% of the price first quoted. Listeners in Singapore point out that "Electronics City", "Sim Lim Tower", and even Shaw Towers (Beach Road) have slightly better prices, though these areas are not so well known to tourists.

Prices in Tokyo, Japan are considerably lower at the "Akihabara" district electronic shops than elsewhere in the country, although they are not as low as they were in 1991. This is also an excellent source for electronic components for constructional projects. If you are leaving the country remember to fill out a TAX-FREE form, and Value Added Tax is immediately deducted.

Conclusions: The best prices are currently found as follows (cheapest first): Singapore (town center - not the airport), Hong Kong (town center - not the airport), Tokyo (Akihabara district).

NORTH AMERICA / EUROPE / AUSTRALASIA

The US Dollar - Japanese Yen exchange rate suffered major changes in the latter part of the 80's, and also for the first two years of this decade. There have been some price increases in North America as a result, although some Japanese manufacturers have absorbed part of the currency fluctuations themselves. Prices in Canada tend to be higher than in the United States. In general: Japanese equipment is between 20-50% cheaper in the Far East and North America, than in Europe. Just compare the prices! This is partly because of EEC import duty which is applied to Japanese equipment, and partly because of a completely different pricing structure.

The restrictions on shortwave coverage on sets sold in the Federal Republic of Germany have now been relaxed, although some shops still stock receivers that only have coverage up to 26100 kHz. Sets sold in Italy, Australia and Saudi Arabia are subject to national restrictions either for economic or military reasons.

THE "GRAY" IMPORTER

If you pick up many electronic periodicals, or the Sunday editions of some national newspapers (e.g. New York Times) you will find quite a few companies offering shortwave receivers with anything up to 30% discount. Based on feedback, we STRONGLY ADVISE you to read the small print in advertisements for shortwave receivers. The "photo" discount centers are particularly popular on the US East Coast (especially New York), but also now seen in Britain, Australia, and the Federal Republic of Germany.

Such companies operate with low overheads selling well-known brand names (SONY, PANASONIC, etc.) at well below the price of authorized dealers. The majority of discount centers do this by buying "gray" market receivers at a discount price in Hong Kong or Singapore. The sets ARE made by the companies they claim to be, but they may offer different specifications than the sets on sale through "official" dealers at a higher price. Check the FM coverage for instance...is it the range used in your country? You would be well advised to check that the discount center really has the set in stock before parting with money via mail-order. Check the details of the guarantee, as some dealers of Japanese equipment in North America levy extra repair charges on sets types not bought through "official" dealers. The national appointed dealers say that the "gray" market is cheating...the national dealership have to invest time and money in publicity campaigns and they cannot recoup that money from the gray importers. In general, modern shortwave receivers are reliable but if they do go wrong it is unlikely that you would be able to fix it yourself. We leave you to decide whether the price difference between "discount" and "authorized" outlets is worth the security of a limited guarantee and back-up service. Most portable radios are marketed to last about 3 - 4 years. In practice, with careful handling, they will last much longer.

THE "ONE-MAN-BAND" SHORTWAVE CENTERS

For the last fifteen years quite a few shortwave listeners and/or engineers have set themselves up in business to supply shortwave receivers to the general public. The number of people doing this has dropped in the last five years. Some offer discounts, others also offer modifications, usually filter replacements. Remember that sets modified by such companies may well make the original guarantee from the manufacturer void. Be sure to check the full details in case you have cause to complain. That said though, the majority of such outlets are run as genuine businesses. They have done a tremendous amount to promote both the existence and improvement of shortwave receivers.

THE HAM RADIO FLEA MARKET

Amateur radio meetings and fairs often have a section devoted to secondhand equipment. Some of the largest in the world are held in Dayton Ohio USA and Friedrichshaven Germany each year. If you wander around you'll immediately see the equipment that didn't sell over the last few years...there will be too much of it. You may also see communications receivers on sale. However, before you part with hundreds of dollars on a second-hand set, bear the following points in mind:

*Does the set really work? If it doesn't how are you going to fix it? Is there a full technical manual supplied with the set? Can you understand it and carry out any repairs? Are the spare parts still available?

*Has the set been modified? Any modifications should have been documented and ideally should be reversible. Remember if you want to sell the same set in the future, a potential customer may want the radio without the modifications. Look at the hand work done by the "modifier". Scorched plastic covers and untidy circuit boards are a sure sign that someone has been "fiddling".

In short, there is a gamble buying a second hand shortwave receiver at a flea market. While prices for used equipment may be higher at a reputable dealer, they should offer some security for the extra price and be able to supply documentation and service.

Whilst we appreciate that readers of this book may want to buy receivers for different reasons (e.g. good mediumwave AM performance, or sensitive FM), bear in mind our comments primarily relate to a receiver's SHORTWAVE PERFORMANCE.

HAMRADIO FRIEDRICHSHAVEN

CHAPTER 2
STEPS TO PICKING THE RIGHT RECEIVER

You can spend anything from US$50 to US$50,000 on a shortwave receiver. But price is by no means the only factor to consider. First you should decide what sort of listening you want to try. In Europe, you are unlikely to hear a 10 Watt station in Ecuador on a 50 dollar portable. But conversely it would be foolish to buy a US$5000 communications receiver for simple listening to major international broadcasters (e.g. BBC, Deutsche Welle, Voice of America, Radio Moscow, Radio Netherlands, etc.)

FOUR GENERAL CATEGORIES OF SHORTWAVE RECEIVERS
A) PORTABLES: If you only listen to the stronger international broadcast stations, or if you plan to take a simple receiver on holiday with you, then consider sets in this category. To avoid disappointment, make a note of the following:

These receivers give cheap and cheerful performance, but you must not expect too much. Their size and weight is given where known, but that should not be the only factor to consider

Some sets only offer PARTIAL coverage of the shortwave bands. If you plan to buy such a set to listen to (a) particular station(s), then check in advance that the set will cover the frequencies you want to listen to. On some receivers the new 22 meter broadcast band (i.e. 13 MHz) is MISSING. This is already a serious disadvantage.

Ease of tuning is another factor to consider! The shortwave bands are extremely crowded and, due to the price range, the shortwave bands may be cramped into a few millimeters on the dial. The sets listed here are better than average in this respect, but if you find the job of searching for stations difficult, go for a set that offers "digital readout". This means the set displays the frequency it is tuned to, rather like a digital clock. This eliminates most of the "guess-work" as to where you are in the band. This is rather more difficult with the conventional "point-and-dial" system.

SOME radios do not have the facility for single-sideband (SSB) reception, so you can't listen to amateur radio operators, radio teletype stations, or utility services.

Selectivity (i.e. the ability to separate the station you want from the interference) and dynamic range (see later notes for an explanation of this term) are NOT good enough for picking out very weak stations.

Some receivers offer the Phase Lock Loop (PLL) tuning systems. This means the set is generally more stable and easier to tune than receivers using older techniques (such as Wadley Loop). Sets offering FM (VHF) coverage, portability, built-in clocks, and stereo adaptors for FM performance just add to their price without offering better shortwave reception.

B) SERIOUS LISTENER: If you plan to listen to programs and also search around for much weaker signals as part of a technical interest, this should be the group of sets you're looking for. They offer good sensitivity (ability to receive weak stations) and variable selectivity (ability to separate stations very close together on the dial), well above the requirements for program listening. As a result, these sets can be used for searching for very low-powered stations in Africa, Asia and Latin America. All have stable SSB reception and most are suitable for radio-teletype reception. All will also give excellent results for program listening, usually with better or equal quality to sets in class A. All have digital readout, and PLL synthesizers for tuning. Hence they are very stable. Sometimes with (often minor) modifications made by certain authorized dealers (see addresses at the end of this publication) such receivers can also perform as well as sets in much higher price brackets.

C) SEMI-PROFESSIONAL MONITORING RECEIVERS: If listening to shortwave is part of your business, or you need higher-than-average performance, then the sets in this category should be considered. They offer better resistance to overloading by strong nearby stations. This problem normally shows itself as the appearance of stations on the dial where they should not be and, in fact, are not really broadcasting. They offer "state-of-the-art" features, but you'll have to pay for them. Previous shortwave experience is necessary to make full use of such sets, so they are usually NOT recommended for the first time buyer or casual listener.

D) SHORTWAVE IN THE CAR: Listeners continue to request details as to whether car radios with shortwave exist. 15 years ago two German companies offered a converter that fitted between the antenna and a standard mediumwave (Broadcast Band) radio. This meant that the user had to tune the car radio without the benefit of knowing the exact frequency being received. All converters have now been withdrawn from the world market, though may be available second hand. Car radios with shortwave have made a comeback in the 90's with two new sets on the world market offering better than average results. Note though that interference from car ignition systems rules out all but the stronger signals while driving, and antenna possibilities are limited. None have SSB.

POINTS TO CONSIDER

1. Decide on what you can afford and choose from the highest possible category. You CANNOT expect $2000 performance from a $200 receiver.

2. Before assuming your present receiver is useless, or that it needs upgrading, check your antenna facilities. Do not always assume that a long (10 meters or more) random wire antenna in the garden will work wonders. If the set is in category A, you may find that such an antenna will cause too much signal to be fed into the receiver's sensitive circuitry. This results in overloading and the appearance of strange stations on odd parts of the dial. Consider an antenna tuning unit.

3. NEVER expect that by buying an expensive receiver you will be able to tune in exotic countries with hi-fidelity reception. Shortwave signals have to travel vast distances. The imperfections of the ionosphere that forms part of the signal path lead to fading and distortion. As yet, there is NO SUCH THING AS A "SUPER POWERFUL" receiver that pulls rare stations in with "local" quality. It is far better to start with an inexpensive set, learn about the shortwave bands, propagation and the limitations of your receiver. If you decide that international radio listening is interesting enough to start chasing weaker signals, upgrading is always possible later.

4. NEVER let salesmanship at your local store talk you into buying a receiver without at least a demonstration. There is still a serious problem when it comes to getting advice from shop assistants, many of whom are not aware of the existence of shortwave broadcasting stations. Their main expertise is in video or hi-fi. If the salesperson cannot name a major international broadcasting station you can hear on the set, or fails to give a convincing explanation of terms like SSB, the chances are that he is bluffing! In Europe, Australia and North America quite a few dealers have started specializing in shortwave. Details are given in the final section of this publication.

5. NEVER buy a receiver in the hope that you can buy extra parts at a later stage to upgrade its performance. Some more expensive receivers can have modifications done, but these are the exception rather than the rule. You CANNOT economically make a professional receiver from a US$150 dollar portable!

6. AVOID using the receiver on batteries if there is an option of AC mains electricity. Batteries are up to 1000 times more expensive per unit of electricity than the household current supply.

CHAPTER 3

SORTING OUT RECEIVER SPECIFICATIONS

I f you page through the companion to this book, the World Radio TV Handbook, it is quite clear that both domestic and international use of the shortwave bands is increasing. Reception quality at the receiving end depends on various factors, but probably most important is the type of radio being used. Most of the tuning aids recently put on consumer equipment have made it easier to find a station, but have NOT necessarily improved reception.

For professional monitoring, receivers costing around US$100,000 are not unusual. A reader survey in mid-1992 showed that readers of the World Radio TV Handbook are usually willing to pay between US$100 - 1000, depending on the type of listening he or she wants to do. In this range it certainly pays to check which features are being offered, and to what standard. However, in general, manufacturers don't make direct comparisons easy. There is no standard "table of results" for cross-reference. Specifications such as "dynamic range" are either left out, or quoted without details on how they were measured. A pin-head looks huge if you're measuring it with an electron microscope! This article explains some of the more common terms you will see quoted, and examines their significance.

WHAT TYPE OF LISTENING?

In the consumer field, you can roughly divide international broadcast listening into three types:

A) Occasional listening to major international broadcasters. Keeping up with news from home whilst on a foreign holiday.

B) Active hunting of broadcast stations, especially on the tropical bands of 90 and 60 meters. The receiver needs to offer ways to dig weak signals out of the background noise.

C) The enthusiast who not only actively listens to weak broadcasting stations, but who regularly monitors utility stations. Radio-teletype (RTTY),

FAX, and Telex-Over-Radio signals can only successfully be monitored if certain features are offered on the receiver (like stable single-sideband).

Most "paperback book" size shortwave radios can be put into the class A type. Although some offer what is termed "single sideband", the tuning is often too coarse to allow easy reception of radio-teletype signals or even amateur radio operators. Bandwidth filters are a compromise. If the manufacturer makes the filter too wide, then splash interference from stations operating on nearby frequencies becomes intolerable. Make the filter too narrow, and the station's programming sounds as though it is coming out of Edison's first phonograph. Portable sets are built to be sensitive enough to give acceptable results using a built-in telescopic antenna. Attaching a long external wire antenna to such a set often totally overloads the input circuitry.

On the other hand, buying an expensive communications type receiver just to pick up the more powerful international broadcasters is a waste of money. Communications receivers are better suited to the serious listener keen on exploring the weak signals. A set that can move up or down the dial in a minimum of 100 Hz steps is ideal for the shortwave broadcast band listener. However, if you want to use the same set for radio-teletype or for listening to AM stations in the upper or lower sideband mode, a set that tunes in steps of not more than 10 Hz at a time is needed.

Having decided on the type of listening, let's examine the receiver specifications.

THE COVERAGE QUESTION

A set which only tunes the "official" shortwave bands is not much use now. Many stations (e.g. Radio Netherlands, VOA, BBC, Radio Moscow, SRI, etc.) are using frequencies just outside the official shortwave broadcast bands to reduce interference problems. The 1979 "Radio Regulations" of the International Telecommunications Union have clearly defined the shortwave broadcast bands. The frequencies are shown in the table.

METER BAND	FREQUENCIES kHz
120	2300 - 2495
90	3200 - 3400
75	3900 - 4000
60	4750 - 5060
49	5950 - 6200
41	7100 - 7300
31	9500 - 9775
25	11700 - 11975
19	15100 - 15450
16	17700 - 17900
13	21450 - 21750
11	25600 - 26100

The use of the 11 meter band remains somewhat limited especially as we head towards sunspot minimum in 1996/97. Back in 1987, the ITU in Geneva proposed that the broadcast bands of 31, 25, 19, 16, and 13 meters should be expanded, and a new 22 meter band (some manufacturers / broadcasters call it the 21 meter band) would be created. Because of the problems at the 1987 World Administrative Radio Conference, the proposed expansions of the band did not officially take place in 1989. In practice many stations have already started using the expanded portions under the motto "use it or lose it." From 3 February to 3 March 1992, the ITU organized a World Administrative Radio Conference in the Spanish seaside resort of Torremolinos. The last time there was a major re-allocation of the radio bands was back in 1979. Since then, the use of the spectrum has increased so much that a new allocation of the ether was badly needed.

The main topics of discussion at the conference were: a new band for satellite radio, more space for shortwave broadcasting, a new band for HDTV (high definition TV), more space for mobile communications and new channels for space aviation and future lunar colonies!

The important decisions affecting international radio listeners were that the shortwave broadcast bands will be extended by 790 kHz in future, although the allocations will not officially come into force until the year 2007. These are the extra allocations:

EXTRA ROOM FOR BROADCASTING

METERBAND	FREQUENCIES ALLOWED FOR BROADCASTING	INCREASE
49 M	5900-5950 kHz	(50 kHz EXTRA)
41 M	7300-7350 kHz	(50 kHz EXTRA)
31 M	9400-9500 kHz	(100 kHz EXTRA)
25 M	11600-11650 kHz 12050-12100 kHz	(100 kHz EXTRA)
22 M	13570-13600 AND 13800-13870 kHz	(100 kHz EXTRA)
19 M	15600-15800 kHz	(200 kHz EXTRA)
16 M	17480-17550 kHz	(70 kHz EXTRA)
15 M	18900-19020 kHz	(A NEW BAND)

You can be sure that many broadcasters won't wait until 2007 before they start to appear on these frequencies. In fact some broadcasters, such as WEWN in Alabama, USA, have already started using the 15 meter band. Radio receivers marked with continuous coverage between 1.6 - 30 MHz will be able to tune both old and new bands.

STABILITY

Just because a radio receiver has a digital frequency readout does not imply that once it is set to a chosen frequency it will stay there. Cheap portable radios are often fitted with a free-running oscillator. Most modern communications receivers, however, contain at least two quartz-controlled oscillators. One is used in mixing the received signal to the first intermediate frequency (IF) stage. The other is there for mixing the first into the second intermediate frequency stage. Often a third crystal oscillator is used for single-sideband reception. Such quartz-oscillators are temperature sensitive. Most sets take about 30 - 60 minutes to reach their operating temperature. Transistors and power transformers warm up and, as a result, the receiver drifts off the frequency it was originally set to. Measuring this at the WRTH laboratory is quite simple. The receiver is tuned to one of the standard time and frequency stations operating worldwide on 10 MHz. The set is switched to single sideband and adjusted so that a tone of 400 Hz is produced in the loudspeaker. The frequency of this tone is measured accurately with a frequency counter (also linked to the time & frequency standard). By putting the receiver in a specially designed room where a constant temperature of 20 Celsius can be maintained, measurement of receiver drift to within 0.01 Hz is possible. Frequency drift in the first hour after being switched on is to be expected. After that time, the set should have settled down. Assuming a constant room temperature, many budget synthesized receivers have a stability of about 5×10^{-6} per MHz. So if a receiver is tuned to 10 MHz, it can drift off the desired channel by up to 50 Hz. That is acceptable for broadcast reception, but causes serious problems if you're decoding a Telex-Over-Radio (TOR). That kind of drift means you will have to make regular manual adjustments.

SENSITIVITY

Advertisements continually boast that a radio is "powerful" or "extremely sensitive". Sensitivity is a measure of how strong a signal has to be at the antenna terminals before it can be determined to be "intelligible". It seems simple enough. Many enthusiasts compare receiver brochures using the rule "the more sensitive, the better!" In fact, sensitivity is not as important as you might imagine, especially when the transmitters you are trying to receive are using powers in the order of millions of watts! A receiver that offers a sensitivity of 0.1 microvolts will NOT necessarily pick up more stations that one which offers a figure of 0.5.

Let's examine the definition more closely. Just what sort of signal is regarded as "intelligible"? This is loosely interpreted as a ratio between the desired signal and the background noise. But weak signals are rarely received without any extra noise. So we modify the definition to be the ratio between the desired signal + the background noise, divided by the background noise alone. However, if the receiver also introduces distortion into the received signal, this needs to be considered too. Professional organizations in Europe use the SINAD norm, which is simply the:

$$\frac{\text{signal} + \text{background noise} + \text{distortion}}{\text{background noise} + \text{distortion}}$$

An experienced telegraphist is said to be able to read a Morse (CW) signal which is just 1.4 times (2 dB) stronger than the background noise. For speech, a stronger signal is needed. The SINAD value is around 12 dB, i.e. the signal needs to be 4 times stronger than the background. The speech or music is then readable with considerable difficulty. For moderate intelligibility we need a value of 20 dB (10 times stronger) and for easy listening this figure rises to 26 dB (20 times stronger). The signal is said to be "noise free" if it is 100 times stronger (40 dB) than the background noise. When sensitivity specifications are quoted in this Equipment Buyers guide and in the World Radio TV Handbook the signal+noise/noise ratio is taken to be 10 dB, or 12 dB in the SINAD norm. Speech and music are therefore JUST understandable, as you'd find when listening to weak distant signals (often termed "DXing").

BANDWIDTH INFORMATION

However, if you pick up a brochure and it tells you the radio has a sensitivity of 1 microvolt for a 10 dB S+N/N, you haven't been given the full story. The bandwidth filter used in the measurement needs to be specified.

Components inside the receiver generate their own noise. This background noise is dependent on the intermediate frequency filter selected. Let's say that a chosen receiver, using a 6 kHz IF filter, needs 1 microvolt at the antenna terminals to provide an intelligible signal, in other words 10 dB S+N/N. If a 600 Hz CW filter is now switched in, the sensitivity rises by 10 dB. Only 0.3 microvolts are now needed to give the important 10 dB S+N/N value. And this is the same receiver! So check that the bandwidth is quoted when you're looking at sensitivity measurements.

To make the story complete, the impedance of the antenna terminals should also be mentioned. The values quoted, and measured by the WRTH lab, are correct at 50 Ohms.

MODULATION DEPTH

When standard AM broadcasting is considered, the modulation-depth of the transmitter being received also affects the intelligibility. In general, an overall modulation depth of 30% is assumed when referring to AM broadcast reception. However, this is a rather outdated norm. Indeed, many companies manufacture devices to ensure the average modulation level of most mediumwave transmitters (shortwave broadcasters are still learning!) is around 60-70%. Perhaps the Orban "Optimod" is the most famous device in this field. This "punches" the signal through background noise much better. If you see a figure for AM sensitivity quoted at 60% modulation and want to know what it would be like at 30%, just divide the figure given by two.

This may sound very complex, but in fact there is an over-riding factor that prevents a receiver with a sensitivity of 0.1 microvolts (6 kHz bandwidth, 10 dB S+N/N) receiving any more stations than one offering a figure of 0.5 microvolts (6 kHz bandwidth, 10 dB S+N/N). The reason is atmospheric noise!

ATMOSPHERIC NOISE

An antenna not only picks up radio transmitters but also background noise. There are in fact various types. Galactic noise is one example, the sun being the main source. Up to about 100 MHz, inter-stellar noise is also present. Alongside that we have atmospheric noise. This is partly caused by the approximately 30,000 thunderstorms that occur each day. All the lightning produced adds to the background noise, though the level does depend on where you live. Listeners in the tropical zones of the world suffer much more from this type of interference. The final type of noise is man-made. This is caused by the millions of electrical devices in use at the moment, from washing machines to fluorescent lamps, and from drills to car ignitions. Even if you live out in the country, overhead high-tension power lines act as giant antennas "broadcasting" electrical noise right across the spectrum.

All the electrical interference forms a background of man-made noise which, even in the countryside, is some 10 times stronger than atmospheric noise.

TABLE SHOWING LEVELS OF GALACTIC NOISE IN MICROVOLTS AT 50 OHMS

FREQ IN MHz	BANDWIDTH 2.1 kHz	BANDWIDTH 6 kHz
1.5	6.30	11.00
3.0	2.00	3.50
7.2	0.90	1.60
9.0	0.63	1.10
15.0	0.36	0.63
21.0	0.25	0.44
25.0	0.20	0.35

The table above shows how strong the galactic noise is at the 50 Ohm input terminals of a receiver. This assumes a tuned dipole is used as the antenna. We've calculated this at two commonly used bandwidth filter values. Remember this is only galactic noise, and that man-made noise levels can be anything up to 10 times stronger.

The figures prove that there is little point building a single-sideband receiver with a sensitivity of 0.1 microvolts for use below 25 MHz. Suppose you are looking for signals on the 41 meter broadcast band. A signal, on say 7200 kHz, is producing a signal strength of 0.9 microvolts at the receiving end. It is just as strong as the galactic noise. If you want the signal to be intelligible, (S+N/N = 10 dB) then the transmitter has to be 3.16 times stronger than the galactic noise, i.e. at a level of 2.8 microvolts. This is therefore the relationship between the transmitter level and the galactic noise. The receiver isn't involved at this stage at all! A more sensitive receiver will simply show a higher signal strength reading on the meter than a less sensitive set. But the signal to noise ratio, and thus the intelligibility, REMAINS THE SAME!

It is for this reason that putting a wide-band antenna amplifier at the front-end of a sensitive receiver doesn't usually help at all. You see a rise in signal strength, but that is all. In practice, the figures given are very much worse, simply because we've ignored man-made interference for the purpose of clarity, and assumed that the receiver circuitry itself generates no noise.

So what sort of figures should one look for? In the professional world, one is usually looking for a sensitivity for 10 dB S+N/N around the same figure as that for the galactic noise. The table above therefore gives you the guidelines of what to look for.

There are more reasons for not building a "super sensitive" receiver. It is better to concentrate on reaching high standards in the following specifications; blocking, inter-modulation, cross-modulation, and selectivity.

BLOCKING

Unfortunately, the antenna outside is not only receiving the signal you're trying to hear. Especially with antennas that receive a wide range of frequencies, the entire shortwave spectrum is presented to the receiver's terminals. The strength of the total energy can vary from around 1 microvolt to between 200-500 microvolts. The automatic gain control (AGC) on a receiver is designed to ensure that the fluctuations in received signal strength (due to fading) are compensated for by raising or lowering the gain.

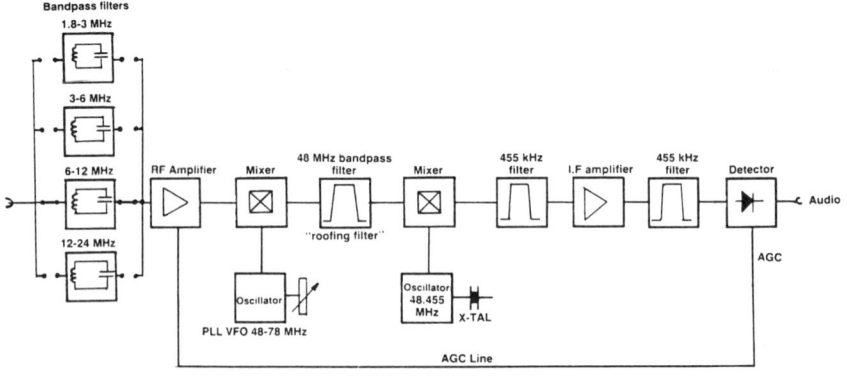

FIGURE ONE : Simplified Block Diagram of a modern receiver.

The above figure shows a simplified block diagram of a modern receiver. It has a 1st intermediate frequency of 48 MHz, and a second IF which is much lower at 455 kHz. Take an example of a station received on 8000 kHz. The signal first goes through a band-filter, is amplified and sent into the first mixer. There it is converted to 48 MHz, and subsequently to 455 kHz by the second IF stage. The selective filters are in the second stage.

The selected filter picks out the desired transmitter from the band between 6 and 12 MHz. The detector is used to create the desired audio, as well as the Automatic Gain Control (AGC) level. If the signal strength is above a preset level, the AGC reduces the gain of the radio-frequency amplifier. This ensures that the signal level to the first IF stage remains constant. It seems a simple and effective system, for in theory only the desired transmitter is allowed to pass the selective filter. In practice though, the section of spectrum that is allowed through the first part of the receiver (in this case 6 - 12 MHz) is taken through the RF amplifier and the 1st IF stage. If you chose to listen to a weak maritime station on 8 MHz, not only is that signal applied to the first mixer stage, but very powerful signals from the 41, 31, and 25 meter broadcast bands. And that's where the problem starts.

Because you've chosen a weak signal, the RF gain is turned up to maximum. It is also amplifying the nearby strong broadcast signals. The first mixer therefore receives those broadcast signals a factor of 3 - 10 stronger than they are being delivered to the input terminals by the antenna. In Europe a simple dipole antenna can deliver signals with a power of 50 - 100 millivolts from the 7 MHz broadcast band without much problem. This level drops a little at higher frequencies, but it is still in the order of 30 - 50 millivolts. If an active antenna is being used, then the signals may be in the order of several hundred millivolts! None of the signals are attenuated by the automatic gain control. Mixers and amplifiers can only handle a certain amount of signal before things go wrong. Weak signals must be above the background noise level to be readable, But at the same time, strong local signals must not be made too strong so that circuits are overloaded.

The results are often audible in two ways. First, a strong transmitter is heard all over the spectrum (in this example between 6-12 MHz), and adjusting the tuning knob does nothing. Or secondly, the signal strength of the desired signal drops, until it disappears in the background noise. The receiver has become insensitive, and this problem is termed "blocking". This is then a measurement of how strong an UNDESIRED station has to be to cause receiving problems to the DESIRED signal.

This problem is true of the simple receiver designs found in many portables. Listening to strong broadcasters is no problem, as the receiver's AGC reduces the gain, and no overloading takes place. But if the receiver brochure mentions a figure for blocking, without saying how it was measured, then it's rather like announcing a production figures in percentages.....percentage of what?

Many laboratories involved in this type of work have adopted measurement norms laid down by the CEPT. This is an umbrella-organization of

European PTTs. This method is designed to test professional receivers for use on board ships. Many of the cheaper amateur/consumer receivers don't reach the required standard, but that doesn't matter. It is important that the procedure is defined. If reviewers and manufacturers all used the same method, the comparison by the consumer would be a lot easier.

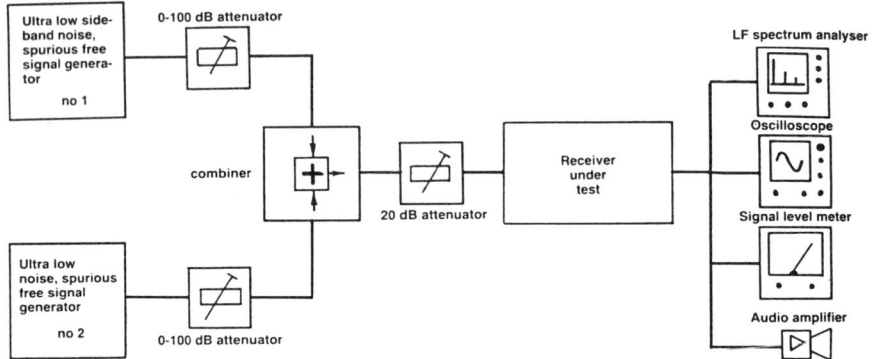

FIG 2. Measuring of blocking, dynamic selectivity, RF protection ratio, image and spurious rejection.

The measuring set-up is shown above. Two high-quality signal generators are needed. They are both stable, offer very low sideband noise, and are free from spurious signals. They feed into a combiner unit, followed by a 20 dB attenuator. This is used to ensure that any changes in the input impedance of the receiver can be compensated for. An audio-frequency level meter and a spectrum analyzer are attached to the receiver's output.

Signal generator A supplies the "wanted" signal. It is modulated and adjusted in level in such as way as to give a S+N/N ratio of 20 dB. The absolute values will of course depend on the sensitivity of the receiver. Signal generator B is unmodulated, and set to operate at a frequency plus or minus 20 kHz from signal generator A. The level of signal generator B is now adjusted so that the S+N/N ratio on the receiver drops back from 20 to 14 dB, or the audio level output drops by 3 dB, whichever occurs first. This level from signal generator B applied to the antenna terminals is therefore the maximum signal the receiver can handle without "blocking". The result may be expressed either as an absolute figure in millivolts, or as a level compared to the wanted signal (in dBs). The CEPT test says that the blocking level should be at least 65 dB (1780 times) higher than the wanted signal. It is quite often the case that consumer/semi-professional receivers don't have sufficient dynamic selectivity to cope with a strong unwanted transmitter using a channel 20 kHz away from the desired signal. This is often the case when a 10 kHz filter is selected when listening to an AM radio station.

PROBLEMS

A blocking level measurement is now made using a frequency difference of plus or minus 200 kHz instead of 20 kHz. It doesn't sound too much different to the first test. But let's look at circuit design costs. In most cases, manufacturers try to make the first mixer perform as well as possible. If budget is limited, then less attention is paid to the second mixer. Remember that stations operating on frequencies sometimes quite distant from the chosen channel are also allowed through the first mixer stage. As long as the strong unwanted signal is outside the frequency range of the 1st mixer stage, then the blocking figure is determined by the 1st mixer stage. But if it falls INSIDE the range, then the blocking is then determined by how much the 2nd IF mixer and amplifier can handle. On cheap receivers the first IF stage is often very wideband. On consumer sets of the 1970's, such as the Barlow Wadley XCR-30 and Yaesu FRG-7, the first IF was 1 MHz wide. Higher class (and as a result more expensive) communications receivers use crystal filters in the first stage to limit what is let through to between 10 and 15 kHz. A good quality narrow-band first IF is particularly noticeable when you're trying to separate two stations a few kilohertz apart, especially when the unwanted signal is much stronger than the desired station.

SELECTIVITY

Selectivity is the ability of a receiver to separate the signal you want from others operating on nearby frequencies. A receiver's selectivity is determined by the quality of the 2nd intermediate frequency filter, usually around 455 kHz. Technically, it's much easier (and cheaper) to make a good filter at 455 kHz, than trying to make a filter to operate at much higher frequencies. There are two important factors that determine such a filter's performance, namely the bandwidth and the "shape factor". In general, mechanical and crystal filters offer better results than ceramic types, though some of the better ceramics perform as well as cheap crystal devices. A good filter is expensive, so at consumer level a compromise needs to be made between performance and price.

Figure 3 shows the response curves of a number of different IF filters measured recently by the WRTH. Note that we've chosen a typical part of the shortwave broadcast bands where transmitters are operating at 5 kHz apart. Transmitter 3 is the desired signal. The better the filter, the steeper the curve, so that the levels of the transmitters on nearby channels are attenuated as much as possible. -6 dB attenuation means that the level is reduced by half, -60 dB means it is 1000 times weaker. In brochures, the bandwidth of the IF filter(s) is usually measured at -6 dB. Some manufacturers measure this from the first -6 dB point to the second. This would give a result of 6 kHz bandwidth in the case of filter 1. Other manufacturers might describe it as Ò 3 kHz, which looks better on paper to the uninitiated. The + or - doesn't mean "about 3 kHz", but that the filter is in fact twice 3 kHz, i.e. 6 kHz wide.

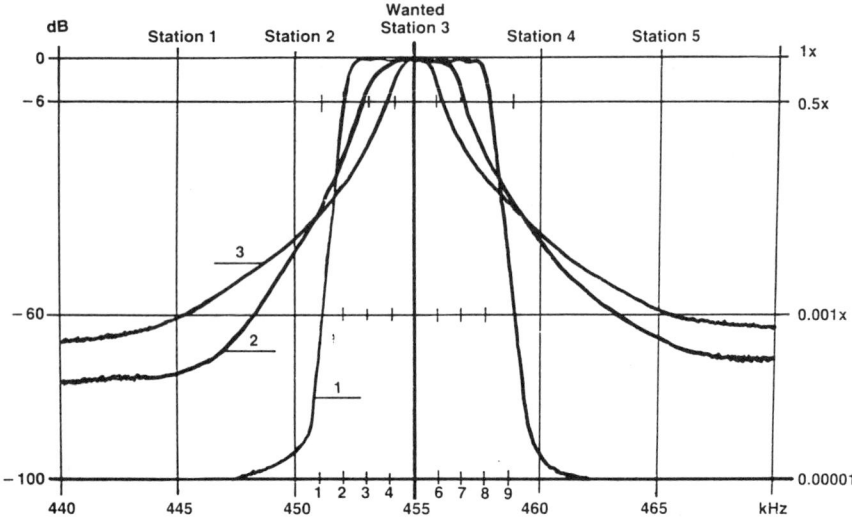

FIG 3: Filter response curves

A single figure for an IF bandwidth though is by no means the full story. As Figure 4 shows, the bandwidth at -6dB for the cheap ceramic filter is 2 kHz. The more expensive crystal filter gives a figure of 6 kHz. But the graph shows that the crystal filter is much better at suppressing unwanted stations on nearby frequencies. The bandwidth at -60 dB is also a useful figure. Many reports also quote the shape factor, this being the ratio between the -60 dB bandwidth measurement and the -6 dB measurement. Crystal filter number 1 in figure 4 gives results of 8 kHz at -60 dB, and 6 kHz at -6 dB. The ratio is 8/6 or 1.33. The nearer the shape factor approaches unity, the better.

SELECTING THE RIGHT BANDWIDTH

So what kind of bandwidths are best? If you plan to listen to local mediumwave AM broadcasters which are separated by 9 or 10 kHz on the dial, then a filter of 9 kHz is ideal. It allows as much of the higher audio frequencies though as is being transmitted, resulting in good fidelity. But the same filter gives disappointing results on shortwave. A 9 kHz filter usually is often as much as 18 kHz wide at the -60 dB point. Using it on the shortwave broadcast bands, where stations are 5 kHz apart, leads to weaker signals being "crowded out" by stronger stations on nearby frequencies.

For general program listening, a -60 dB figure of around 9 kHz is desirable. Depending on the shape-factor, that leads to a -6 dB figure of between 3 and 6 kHz. The wider the filter at -6 dB, the better the fidelity. In theory, shortwave stations don't put out audio frequencies higher than 4500 Hz. In practice though, many stations "forget" the filtering. This leads to audio response up to about 9,000 Hz, which "accidentally" blots out stations 5 kHz either side of the transmitter in question. This is just one of the techniques used to ensure a station "shouts the loudest".

There are situations where fidelity isn't important. If you want to dig out a weak signal from between two strong signals, i.e. such as when DXing, then something like a 2.7 kHz (- 6dB) filter is ideal. If you want to use your receiver for utility DXing, then you need a filter that's just wide enough to accept the highest audio frequency being demodulated. In this case filters of between 2.4 and 2.8 kHz are fine for SSB phone or RTTY work. If you plan to listen to a lot of Morse (CW), a filter of 1, 0.5 or even 0.3 kHz may be wide enough.

Only more expensive shortwave receivers offer the listener a choice of bandwidth filters. With the crowded shortwave broadcast bands it is often worth paying a little extra to have at least a second, narrower filter choice. Then if interference gets bad while you're listening to a program, you can switch to the narrower filter to continue listening. The audio quality, however, will become more "muffled".

DYNAMIC SELECTIVITY

It is interesting to see how the bandwidth of an IF filter is measured. It is usually done by applying an unmodulated signal to the filter, and measuring how the carrier level drops as the tuning knob is moved off the center frequency. When you're tuning the overcrowded shortwave broadcast bands however, the situation is rather different. A reasonable IF filter can be expected to attenuate signals outside its bandwidth by some 90 dB. But in order to keep costs to a minimum, many manufacturers solder the chosen IF filter directly onto the printed-circuit board. This leads to some signal leakage around the filter itself, and the 90 dB attenuation is often reduced to between 50 and 60 dB. This is certainly the case when cheap ceramic filters are monitored.

Circuit layout is also important too. If the 1st and 2nd IF stages are not adequately screened from each other, then signals may bypass the filter stage due to coupling effects. If you're trying to build a receiver for the US$500 market, then you cannot afford the type of screening seen on professional receivers of US$50,000.

This leakage problem means that many consumer/semi-professional communications receivers have much poorer selectivity than the static selectivity figures would first indicate. For this reason, it is better to measure a receiver's dynamic selectivity. This is a much better test of how well the set can separate signals in practice.

The measuring technique is identical to that already described for "blocking". Two signal generators are used. One is set to a shortwave broadcast frequency and adjusted to so that the signal+noise/noise level is 20 dB. The second generator is modulated with a 400 Hz tone with 30% modulation. The frequency of generator B is adjusted to move in 100 Hz steps both above and below the center frequency to which the receiver is tuned. The level of the unwanted signal (coming from generator B) is adjusted each time so that the level of the desired signal drops from 20 to 14 dB (i.e. fair interference). The difference in level between the wanted and unwanted signals (given in dB) is therefore a measurement of how strong an signal on a specified nearby channel has to be before you get reception problems.

The European CEPT organization has come up with guidelines for professional receivers for dynamic selectivity.

DYNAMIC SELECTIVITY AM RECEPTION CEPT NORM	
INTERFERING TRANSMITTER	MINIMUM ATTENUATION NEEDED AT:
+ AND - 10 kHz	40 DB
+ AND - 20 kHz	50 DB

DYNAMIC SELECTIVITY SSB RECEPTION (2.8 kHz -6 DB FILTER)	
INTERFERING TRANSMITTER	MINIMUM ATTENUATION NEEDED AT:
- 1 kHz AND + 4 kHz	40 DB
- 2 kHz AND + 5 kHz	50 DB
- 5 kHz AND + 8 kHz	60 DB

In fact the CEPT norm for AM doesn't turn out to be all that stringent. When measuring in the shortwave broadcast band, where stations are only 5 kHz apart, it is assumed the transmitter is only 30% modulated by a 400 Hz tone. Using a 1 kHz tone, and 60% modulation would be far more realistic. In that case the accepted levels would be lower.

RF PROTECTION RATIO

As we've seen, dynamic selectivity and blocking measurements are roughly similar. In both cases we're interested in the level of the interfering signal at a determined spacing when it interferes with the selected frequency. With blocking we're interested in transmitters from about 30 kHz to several MHz away from the chosen channel. With dynamic selectivity, the spacing is just a few kHz. You can also present the information by varying the frequency of the interfering signal between 0 and 100 kHz from the desired channel. We've drawn such a curve below.

We call this curve the RF protection ratio. We've only shown one half of the curve, because in practice the other half is a mirror image. If it is not the same, then the worst curve is chosen. The left-hand vertical scale represents decibels, where 0 dB is equivalent to the sensitivity of the receiver for 20 dB S+N/N or SINAD. The curve then shows how well unwanted signals are suppressed. The right-hand vertical column shows how many times stronger the unwanted signal has to be to cause problems. 60 dB is equivalent to 1000 times. If the chosen receiver has a sensitivity of 2 microvolts for 20 dB S+N/N, then moderate interference (20 dB S+N/N drops to 14 dB) occurs when the unwanted transmitter is 1000 times as strong, i.e. 2

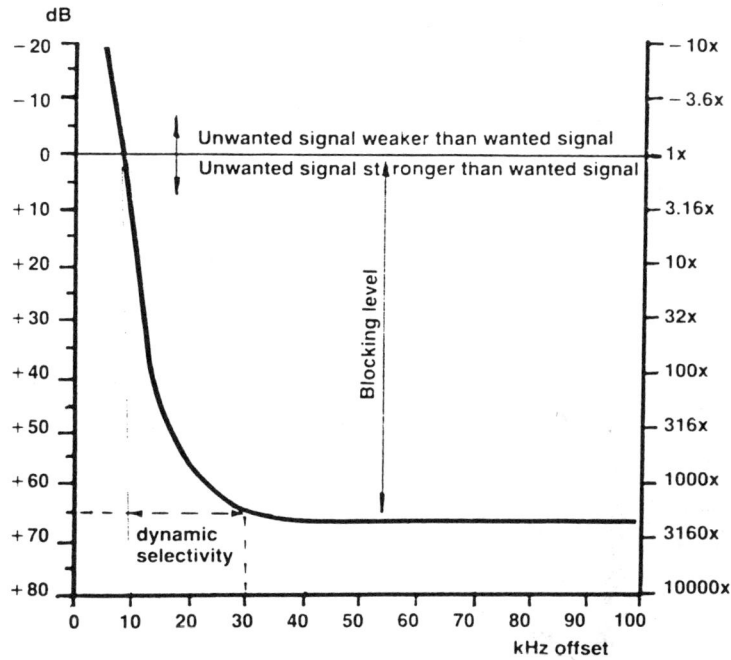

FIGURE 4: RF Protection Ratio

millivolts. The curve shows that further than 50 kHz, the attenuation remains constant around -66 dB. The blocking level is therefore 66 dB = 2000 x 2 microvolts = 4 millivolts. Below 10 kHz distance the receiver is much less able to cope with strong nearby signals. This is to be expected, as the unwanted signal is also being allowed through the IF filter. This shows that the RF protection ratio is a useful way of describing how a receiver can cope with unwanted signals.

CROSS MODULATION

Cross modulation arises when the modulation of transmitter to which the receiver is NOT tuned, appears on top of a signal which is being listened to. The effect is easy to spot. As you're listening to a program you notice that in the pauses of speech or music, another program is heard in the background. Schedules show that the interfering program is in fact not operating on that channel, but further down the band. Cross modulation can affect reception of both strong and weak signals. Its measurement is similar to blocking. The signal generator is modulated at 30% with a 1000 kHz tone. Adjustments are made so that 1 millivolt at the RF terminals gives 50 mW at the 8 Ohm loudspeaker terminals. The second generator, representing an interfering signal, is also modulated at 30% but with a 400 Hz tone. Its frequency is 20 kHz away from the channel chosen by generator A. The second generator is now set to a level that produces a 400 Hz tone in the receiver's loudspeaker. The 400 Hz tone is not the same level though, but 32 times (30 dB) weaker. This is the cross-modulation level, and CEPT standards say that it should not be lower than 30 millivolts. That sounds a lot, but in practice that means it is only 30 dB more than the level of the signal you're trying to receive. Referring back to the dynamic selectivity table, the CEPT says the minimum selectivity (frequency spacing 20 kHz) must be 40 dB.

You can see that a receiver will show signs of cross-modulation before the point of insufficient selectivity is reached. This is especially so if you live close to a mediumwave AM station. But because such an effect is dependent on the power of the unwanted signal, an adjustable attenuator at the front end of the receiver can help. Adding 6 dB of attenuation, for instance, reduces the wanted signal by 1 "S" point on the meter (i.e. by half), whereas the cross-modulation level drops by 2 "S" points, or is reduced to a quarter.

INTERMODULATION

Whenever at least two signals are applied to the front end of a receiver there is always some degree of mixing between the two. This leads to new, undesirable signals. These "intermodulation" products are generated by the receiver itself, and can impair the quality of real signals. First, we have so-called "2nd order" products, and these occur at the sum of both signals, and also at the difference.

Suppose you receive two strong mediumwave AM signals on 1200 and 1000 kHz. The second order products might be generated on 1200 + 1000 = 2200 kHz, and 1200 - 1000 = 200 kHz. Because your antenna isn't only receiving just two transmitters you can see that unsuppressed 2nd order products will cause problems to reception of signals in the longwave broadcast band, and in the 160 m marine band. Hang a longwire antenna on the front of a very cheap receiver and then tune the long-wave band. If you

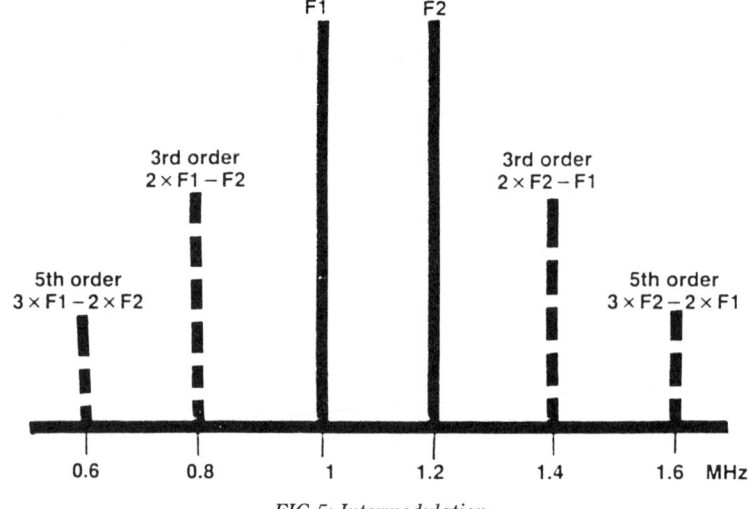

FIG 5: Intermodulation.

hear whistles and distorted broadcast signals on channels where they're not listed, chances are high that the receiver is producing 2nd order intermodulation products.

Better communications receivers, and even some portables, use input filters after the antenna terminals. So if you've selected longwave, only signals between 100 and 400 kHz are allowed through to the first mixing stage. As you tune up the dial, the input filters are switched to allow the portion 400 - 1800 kHz, then 1800 - 4000 kHz and so on. If you're listening to the mediumwave band and hear strange signals, this is not due to 2nd order products, but the 3rd and sometimes even the 5th order products.

The situation is shown in Figure 5. Let's keep to our example of F1 and F2 being 1000 and 1200 kHz respectively. The 3rd order products (mainly occurring in the 1st mixer stage) would appear on 2 x F1 - F2 and 2 x F2 - F1, i.e. on 800 and 1400 kHz. The 5th order products occur at 3 x F1 - 2 x F2 and 3 x F2 - 2 x F1, i.e. on 600 kHz and 1600 kHz. You can see that these products could potentially disturb reception of signals in the mediumwave band.

Remember though that the antenna is delivering hundreds, may be thousands of signals to the receiver's terminals. All these intermodulation products therefore mix together to produce a background noise, preventing reception of weaker genuine signals.

In our opinion, this is one of the major problems of consumer/semi-professional receivers today. Not only is there an increasing number of high-power stations on the air, but modern receivers have simplified their tuning by not including a receiver "preselector". Such a device is still available, sometimes as an optional extra, and helps to reduce the amount of unwanted signals at the first mixer stage of the receiver.

It is instructive to compare a professional receiver (something in the US$10,000) range with a "consumer" communications set costing under US$1000. Tuning between the broadcast bands on the budget receiver, there always appears to be a moderate level of intermodulation noise, whilst the professional receiver is much quieter. Of course the problem with intermodulation is more intense in densely populated (transmitter wise!) Europe. Tests in the Pacific show a cheaper communications receiver to be much quieter there. But even so, the re-launching of good preselectors for the amateur market would solve a lot of listeners complaints.

Using the same double signal generator technique, with signals spaced 30 kHz apart, it's possible to measure the 3rd order intermodulation products. Professional CEPT norms say that the intermodulation product should be set to have a level of 32 microvolts, therefore producing a signal of 50 milliwatts in the low-frequency spectrum. If that is the case, then both undesirable transmitters should not be lower in level than 100 millivolts. Most amateur receivers have long gone into a totally blocked situation before this level is reached. So we have to change the norm. Usually we check how strong the two unwanted signals have to be before intermodulation products of 1 and 10 microvolts appear.

DYNAMIC RANGE

This is the "buzz" term of the last nine years. But what does it mean? Following on from our discussion of intermodulation, we can now set the level of the unwanted 3rd order product so that it is 3 dB above the receiver's own noise floor. This is just detectable on the receiver. As can be seen in Figure 6, the difference in level (given in dBs) between this 3rd order product level, and the strength of one of the two received signals which causes the intermodulation product is the "intermodulation free usable dynamic range". The larger this is, the better.

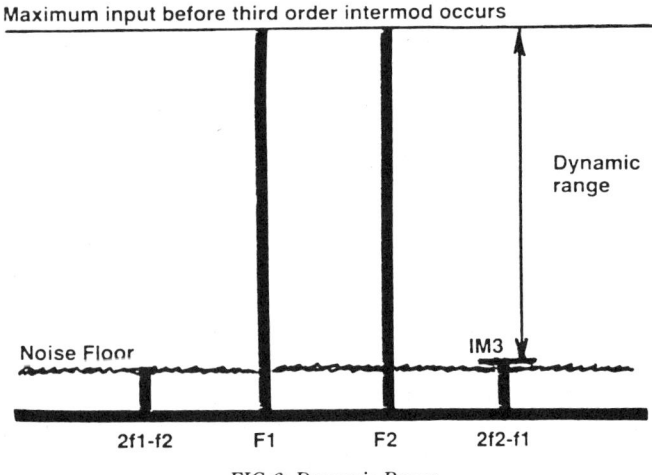

FIG 6: Dynamic Range

Because the potential customer is interested in big numbers, a lot of manufacturers have discovered ways to make the figures in their brochures much better. They're not telling lies of course, but they're not saying how they got the numbers either.

Rather than taking the level of one of the signals, you can take the sum of both of them. This gives you double the dynamic range, in other words, the dB's rating rises by 6 dB. If you see the term Peak Envelope Power (PEP) by the side of the dynamic range figure, then the sum "trick" has been used.

More common is a dynamic range measurement made with a very narrow bandwidth filter. The noise floor drops, and the sensitivity rises, as the IF filter DECREASES in bandwidth. So the smaller the filter, the better the dynamic range. Even famous name manufacturers quote a figure of, for example, 103 dB dynamic range, forgetting to add that they used a 500 Hz CW filter to measure it.

When professional receivers quote a dynamic range of just above 100 dB using the 2.7 kHz SSB filter, it seems suspicious when a set costing one tenth the price comes up with a figure of 106 dB (twice as good).

Conversion of "hyped" figures into "true comparative" figures is possible, though if the manufacturer isn't being specific, neither is the end result. Take off 6 dB if the amplitude of both signals (see Figure 7) has been taken. Assuming a 500 Hz CW filter has been used, you can also take off 6 dB to get a figure for its performance with a narrow 2.7 SSB filter, and another 5 dB for a 6 kHz AM filter.

Budget receivers then produce figures in the range 60 - 80 dB, whilst better sets offer between 80 and 100 dB. Remember every 6 dB rise means a dynamic range which is twice as good!

INTERCEPT POINT

Quick calculations show that the 3rd order intermodulation products are linked to the strength of the original two transmitters. If both of these rise in strength by 10 dB (e.g. because of switching to another antenna), the 3rd order products rise by 30 dB. The reverse is also true, hence again the value of a preselector and switchable attenuator. Although it may sound strange, switching in a few dB's attenuation may be what's needed when chasing weak DX signals. A unit that can introduce 0, 3, 6, 10, and 20 dB is a worthwhile investment, and more useful than the "all-or-nothing" 20 or 30 dB attenuators seen on cheaper communications sets.

The rapid increase in 3rd order products against the slower increase in signal strength is illustrated in Figure 7. Note that the strength of the original signal is given in dBm, i.e. dB's above one millivolt. To help conversion of dBm into something that's easier to grasp, we've shown the antenna voltage needed, assuming a receiver with a 50 Ohm input impedance. 0 dBm is therefore the same as 224 millivolts. On the vertical scale is output signal level (in dBm) after the first mixer stage of the receiver. Amplification is assumed to be unity in this example, so -60 dBm input gives -60 dBm output. Above 0 dBm though, the output signal no longer rises with increased input level. The 1st mixer stage has reached its maximum output.

The small steeper line shows the growth of 3rd order intermodulation products. This is interesting. Draw a line at an input signal of -20 dB, and you can see the difference between the level of the wanted signal and the 3rd order product is 46 dB. Suppose you're trying to listen to an amateur radio operator on around 7050 kHz. A little further up the dial are broadcast stations on 7200 kHz producing signals of at least 20 millivolts (around -20 dBm). Figure 8 shows that 3rd order product is 46 dB (about 200 times) lower than the level of the broadcasts transmitters in the 41 meter band. But this means we're talking about levels of 20 mV / 200 = 100 microvolts. That's a significant background noise!

Rather than draw Figure 7 each time though, a useful trick has been developed. By extrapolating both lines until they intercept, we can determine the "intercept point". You'll often see this quoted. In the case of Figure 8, the value is +5 dBm. Using this figure it is possible to reconstruct Figure 8 if desired. The higher the 3rd order intercept point, the better!

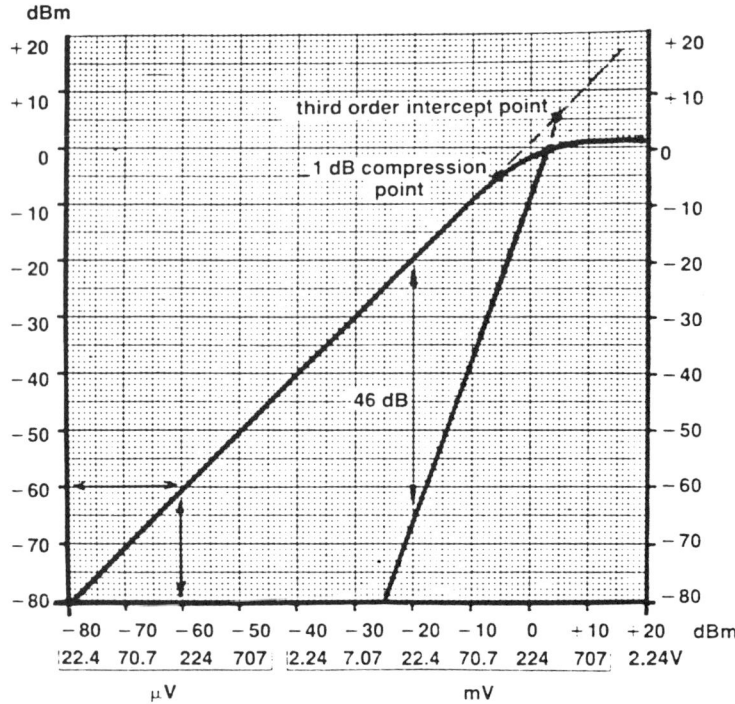

Figure 7: Intercept Point

Cheaper receivers have values between -20 up to 0 dBm (22 mV to 224 mV) whilst the more modern communications receivers offer values between 0 and +20dBm (i.e. 224 mV to 2.2 volts).

Treat an intercept point with some care though. It is meaningless if you don't know whether the front end of the radio has some sort of HF selectivity. Older receivers may have lower intercept points than modern sets. But, especially the military surplus sets have tuned preselectors which lead to much quieter noise-free reception than modern receivers with an octave filter in the input stage. Receivers with a broadband input (in other words, wide open to any signals between 0.1-30 MHz) need an intercept point of at least 40 dBm to remain free of intermodulation noise. Few can reach that level.

CONCLUSIONS

This is an equipment guide, not an engineering manual. So there are some aspects we have left out, like a discussion of AGC time constants. In practice many manufacturers are not giving the customer all the data they need to make a proper comparison. Whilst a specification sheet is a bit too much to ask of a radio costing US$50, we think that any radio over US$300 should come complete with a specifications list.

For the listener the bottom line is simple. Too much signal from the antenna is as bad as too little, especially if you're listening in an area where high power stations on mediumwave are active. If you suffer from these problems, consider the purchase of a good preselector.

There is a disturbing trend in receiver design that we feel is worth pointing out. If you said to someone 10 years ago that a shortwave receiver with digital frequency readout at price of around 200 US dollars was on the way, they probably would not have believed you. It is only since 1980 when Sony Corporation of Japan launched their ICF2001 that tuning of the shortwave bands has become easier, for those who can afford the facility. But thirteen years later, the companies in the business of making some shortwave portables seem to be taking a new direction. This appears to be the route of unnecessary gimmicks, rather than in the direction of improving the radio receiver part of their models.

The market for shortwave portable receivers is not large in most developed parts of the world. In Australia, for instance, one large European company estimates that just 5% of the radios now sold each year have some sort of shortwave coverage. This is needed for people in the outback who want to listen to the domestic shortwave service in the Northern Territory. That means that these types of radio don't get priority in terms of marketing and development budgets.

CHAPTER 4
HARD FACTS ABOUT ACTIVE AND PASSIVE ANTENNAS

Whereas most medium and shortwave listeners would like to have giant rotatable log periodics or beverage antennas in the garden, the real world is different. More and more city dwellers are discovering the convenience of a push-button portable. But our 1993 reader survey confirms earlier WRTH studies that show local authorities are becoming more restrictive in their attitude towards outdoor antennas. Pick up almost any electronics magazine and you'll find someone advertising a solution - the "active" antenna. It is small, easy to install, and its box of electronics means that it performs as well as any "passive" outdoor antenna. That's the claim at any rate.

In fact, any conductor of electricity, be it a lamp post, an iron bar in the wall, or a piece of wire is capable of converting electromagnetic energy, i.e. radio waves, back into an electric current. If such a conductor comes in contact with the ground, energy flows to earth. If we isolate this conductor above the ground, the energy can be fed into a receiver, and signals with a certain frequency selected for their entertainment value. Sounds simple. But note that we're talking about ENERGY not about potential or voltage. That can only occur when an energy source (the antenna) is connected to a load (the receiver). Every energy source has an internal resistance (or more accurately impedance), and maximum energy transfer will occur when this is the same as that of the receiver. It is for that reason that most receivers offer a 50 Ohm impedance input, so as to match 50 Ohm coaxial antenna cable. In theory the antenna and any switches associated with it should also be a constant 50 Ohm too. In practice there is no antenna that always has a 50 Ohm impedance throughout the long, medium and shortwave range. Its impedance is dependent on a number of factors, including its length, height, and shape.

Experimentation with antennas has existed since radio began. One of the basic antenna forms is a dipole, whose length is half that of the wavelength it is to receive. The center of the dipole is cut so that a coaxial cable can be connected. At this length the antenna impedance best matches that of the connecting cable, namely 50 Ohms, and there is a maximum transfer of energy.

A dipole performs best on the wavelength for which it was cut. On other frequencies, reception will also be possible, but the impedance is then much higher than 50 Ohms, e.g. 1000 Ohms, or much lower, e.g. 5 Ohms. This leads to much poorer energy transfer.

There are many solutions to help reduce this problem of mismatch. A longwire, 5 times the length of the desired wavelength, has a more constant but higher impedance. Some communication receivers even have a 600 Ohm antenna connection to suit this type of antenna. But because coaxial cable cannot be used as the antenna lead in wire, such a longwire antenna is more susceptible to local interference sources (fluorescent lights, thermostats, etc.).

Antenna tuners are often advertised as a way to match the antenna to the receiver. But they work best when they're mounted as close as possible to, or even in the antenna. There is little point putting one at the end of a 50 Ohm coaxial lead-in wire - mismatch has already taken place.

There is another way to solve the impedance matching problem.

ACTIVE ANTENNAS

Active antennas solve a number of problems immediately. Few people have the space to hang a dipole of some 30 meters to match their interest in the 60 meter tropical broadcast band. Active antennas often take the form of a telescopic whip which can be unobtrusively mounted on a outside wall or balcony. Such an small antenna has a high internal resistance. In fact it consists of a real resistor (usually a few Ohms, and therefore rather insignificant) and a capacitor. For a whip antenna, this usually has a value around 12 pF. The capacitor offers an apparent resistance, whose value is frequency dependent. At a frequency of 200 kHz this is of the order of 55,000 Ohms, and at 30 MHz around 230 Ohms. In the longwave part of the spectrum, the impedance is quite high. It is useless trying to hang a 50 Ohm coaxial cable to it, for in effect you're short-circuiting it. The trick in an active antenna is to connect the whip to an amplifier with a high-impedance input. The output impedance of the amplifier is designed to be 50 Ohms, and a coaxial cable matches perfectly.

But won't a small antenna receive much less signal than a large one? You'd be surprised. Theoretically, the smallest antenna you could design is a point, i.e. an isotropic antenna. Say this receives an energy from the electromagnetic spectrum equivalent to 0 dB. An ordinary half-wave dipole antenna receives 2.5 dB (or 1.6 times) more energy than an isotropic antenna. If you were to use a small active antenna, such as the DATONG AD270, this gives 1.76 dB more than an isotropic antenna, or 1.5 times the energy. The difference between the standard and small active antenna elements is therefore just 0.39 dB, insignificant in fact. But unlike a small antenna, a large aerial can be directly matched to a 50 Ohm cable, which is why an active antenna uses an electrical solution to change the impedance.

Most active antenna manufacturers add some form of amplification to their impedance matching unit. The active antenna itself may take many forms such as a miniature dipole, a loop, or some type of whip. But there are some disadvantages in using an active antenna!

PROBLEMS

The weak point of most active antennas is the impedance matching unit and/or the amplifier. Many factors affect performance. We noted earlier that a whip antenna has an apparent capacity of around 12 pF. It is therefore up to the manufacturer to design a circuit which has a very low input capacity. Easier said than done. Budget antennas which don't heed to this point have a noticeably lower sensitivity, i.e. efficiency, above about 10 MHz. Higher priced antennas have usually solved this problem.

A far more serious drawback in most active antennas is ability to handle strong signals - intermodulation products.

INTERMODULATION

Active antennas are designed to be broadband. All signals between 10 kHz to 30 MHz are presented at the antenna output terminals. In simple designs, intermodulation products can occur in the antenna's electronic impedance matching unit. Suppose there are two strong local mediumwave transmitters receivable in your area. Take 747 kHz and 1008 kHz as examples. Second order intermodulation products could occur at 1008 - 747 = 261 kHz and 1008 + 747 kHz = 1755 kHz. If there are a lot of strong mediumwave signals available, then check the range 100 - 400 kHz. If longwave reception on the receiver is masked by noise and whistles, the active antenna may be producing a number of intermodulation products. This overloading problem also applies to the shortwave broadcast bands, especially the overcrowded 6, 7, 9, and 11 MHz bands. Mixing from stations on e.g. 7200 and 11800 kHz give products at 19000 and 4500 kHz. The latter may well mask reception of weak 60 m tropical band broadcasters with an unacceptable noise level. And we've only considered 2nd order products in this example. Third-order products may also be present.

In the section on deciphering receiver specifications we've dealt with intermodulation and the intercept point in more detail. The higher the intercept point for the antenna, the more signal it can cope with before unwanted "ghost" signals start appearing all over the dial. Because no preselection in the antenna is incorporated, an intercept point of between 36-40 dBm is needed in Europe to keep intermodulation to an acceptable level. With the exception of the DX-1 from RF Systems and the Liniplex loop antenna, none of the consumer active antennas offer this level of performance. In other parts of the world, where signal levels are lower, problems with intermodulation are less severe. But we're not finished yet!

HARMONIC DISTORTION AFFECTS MEDIUMWAVE AND MARINE BAND.

The better manufacturers do quote an intercept point in their specifications sheet. However, most fail to quote a figure for harmonic distortion. But it is this factor that makes listening to the marine band using many active antennas simply impossible. In the Hi-Fi world we're used to distortion figures of 0.01%. In RF amplifiers, the results are somewhat different. It is very expensive to make an RF amplifier with a distortion less than 1%. It doesn't sound much, but such an amplifier produces a 2nd harmonic which is only 40 dB weaker than the incoming signal. 40 dB attenuation is equivalent to 7 "S" points, which means that any signal stronger than S7 will be audible on its 2nd harmonic. This means a bed of noise and/or strange signals between 1070 - 3200 kHz. The same argument applies for the high-power broadcast bands, e.g. 7100 - 7300 kHz is able to cause a 2nd harmonic area between 14200 - 14600 kHz.

In the worst case then, an active antenna could ensure that the entire spectrum between 10 - 30000 kHz is full with intermodulation and harmonic distortion products. It isn't quite that bad. Any signal which is at least three times stronger than the intermodulation product generated within the antenna will give a signal-to-noise ratio of 10 dB, i.e. just readable. But weaker signals will be buried in the noise. The WRTH tests show that on most of the cheaper active antennas, all strong signals are made even stronger (compared to a passive dipole), but that weaker signals disappear. Because many powerful transmitters are continually varying in strength (due to fading, etc.), so does the background noise due to intermodulation effects. This certainly shows up when you find a rather quiet area of the shortwave band.

POLARIZATION

Whereas polarization is important at VHF/UHF frequencies, it becomes less critical below 30 MHz. Even if a signal starts out as a horizontally polarized waveform, the effects of the ionosphere and the curvature of the earth cannot guarantee the polarization at the receiving end. In fact it is constantly changing, as is the ionosphere. The antenna therefore receives both horizontally and vertically polarized signals, plus those some way in between

This effect is also noticeable with active antennas. Some signals were received better on the Datong AD270 (with its horizontal miniature dipole construction) than on the Dressler ARA 60 (with its vertical whip), and vice-versa. On long and mediumwave, where the wavelength is huge in comparison with the size of the antenna element, this effect is difficult to determine. But above 5 MHz, our tests showed that polarization made a signal strength difference of up to 3 "S" points (18 dB). A signal was just audible on one antenna, and unreadable on the other, depending again what polarization was arriving.

The serious DXer should ideally have two active antennas, one vertically, the other horizontally polarized. A coax switch can then be used to select between the two. Another possibility is the multi-polarization function of an antenna such as the RF Systems DX-1.

ANTENNA DIAGRAM AND INTERFERENCE SUSCEPTIBILITY

Because active antennas are small in comparison to the received wavelength, they are generally sensitive to signals from all directions (omnidirectional). That even applies to the miniature dipole on the DATONG AD270. Only above around 15 MHz does the DATONG exhibit some directional characteristics.

As well as looking at sensitivity in the horizontal plane, the vertical plane is worth considering too. When considering full-size passive examples it can be shown that vertical antennas are better for the reception of signals with a low elevation angle (i.e. coming in from a long distance), whilst a dipole is better suited to signals with higher elevation angles, i.e. those that have travelled via 1-2 hops over the ionosphere.

At low frequencies this effect with the small dimensions of an active antenna is difficult to show. Between about 10 and 15 MHz however, the horizontal DATONG antenna does indeed deliver stronger 1-hop signals than a vertical, such as the DRESSLER ARA-60. These were not spectacular differences. The RF SYSTEMS DX-1 is the only antenna that has the advantage of both vertical and horizontal elements.

The horizontal DATONG antenna exhibits a significantly lower susceptibility to local man-made interference, such as thermostats, electric motors, transformer arcing, fluorescent tubes, etc. In general, this type of interference appears to be vertically polarized. If you live in a city area, then the horizontal DATONG does indeed give a much lower background noise than its competitors.

Active antennas exhibit a specific interference problem. Because they can receive signals at very low frequencies, they are particularly sensitive to the interference signals produced by domestic TV sets. At night, when TV sets are on, the noise level (as a result of the 15.625 kHz TV line frequency, and harmonics) rises by about 6 dB or 1 "S" point on the average European housing estate. Home computers and their associated monitors become a serious source of interference, even if they are some 10's of meters away. Especially in areas where the household power supply comes via overhead cables, the power socket brings not only power, but also all types of low-frequency interference.

Unless you live on a secluded farm away from interference sources (in which case you'd probably have space for a passive outdoor antenna), an indoor active antenna will probably disappoint you. The YAESU FRA-7700, for instance, is extremely sensitive to interference sources in the house, and weak radio signals are often screened from the whip by any iron in the reinforced concrete walls of the house. Indoor loop antennas are less sensitive to this type of interference, because they react to the magnetic rather than the electrical component of the incoming electromagnetic wave. If man-made interference proves to be troublesome, it is a good idea to try different places in the house, even some 1.5 meters above the roof if this is permitted.

OUTPUT VOLTAGE

The output voltage of many active antennas is rather high in many cases, especially as many incorporate some sort of extra amplification. Levels of around 300 mV were measured in Europe from local mediumwave AM stations. Few receivers can handle such a strong signal without causing intermodulation products and overloading. Not all active antennas are fitted with an attenuator, presumably assuming this is available on the receiver. This will be less of a problem in areas where signal strengths are lower (e.g. in the Pacific), but an attenuator is always a good investment.

FADING

This important factor is rarely mentioned. If you're interested in utility signals (Morse, telex, fax, etc.) as well as broadcasting stations, fading causes corruption of the received data. There are two types: fading due to varying signal strength, and multi-path fading. The ionosphere is far from being a perfect refracting medium, and it is in constant motion. As a result, signal strength at the receiving end is constantly varying. Apart from a total loss in strength, minor variations can be compensated for by the receiver's automatic gain control.

As its name implies, multi-path fading is caused by signals that have travelled by different routes. Although the transmitting antenna generates one wave, this may be refracted by different parts of the ionosphere, leading to several signals with the same program appearing at the antenna at slightly different times. Although a fraction of second apart, these waves interact, sometimes adding together, sometimes canceling each other out. This type of fading not only affects the signal strength, but the modulation too. Broadcast signals get a characteristic rolling distortion, whilst telex signals lose an important mark or space leading to data corruption.

But whilst a multipath fade happens at a particular instance at point A, a few meters away reception is normal, the selective fade hitting point B at a different instant. Professional installations can reduce this effect by mounting two antennas some meters apart, and selecting the best antenna at a given time.

It is clear from our extensive tests that active antennas exhibit far more multipath fading problems than full-size passive antennas. A long-wire in the garden is large enough so that there is always some part of the incoming electromagnetic wave which is not canceled out during a selective fade. This is not the case with physically smaller active antennas. The DATONG AD370 and RF SYSTEMS DX-1 exhibit less problems in this respect, due to their larger size. Remember this problem is particularly noticeable when trying to copy utility stations.

THE MEASUREMENTS

11 commercial antennas for the shortwave listener were tested: 4 passive, 4 active antennas designed for outdoor mounting, and 3 for use inside. One way of measuring each antenna is to check the intercept point over the entire frequency coverage. We did this in some, but not all cases. A few of the antennas sealed the electronics in such a way that it was impossible to isolate the amplifier for reliable measurements without damaging the unit. However, as we've seen, antenna performance is also highly dependent on the receiver being used. We decided therefore that the core of the test would consist of extensive practical comparisons.

It is easy to jump to conclusions with active antennas. Compare one with a longwire antenna in the garden, and the active antenna usually delivers a much stronger signal to the receiver. This doesn't mean to say it is better. The active antenna may be amplifying atmospheric noise, and generating a whole range of intermodulation products.

In 1987 the World Radio TV Handbook made a special series of antenna tests which went down well with readers. In June and July 1992 we went back to the test site in The Netherlands, located in the countryside on the outskirts of a small town in the east of the country. Man-made background noise is measurable, but not at a high enough level to block weak HF signals. The nearest mediumwave transmitters (2 x 400 kW) are 60 kilometers away. The passive antennas (dipole, Sloper, and T2FD) were all mounted 10 meters above the ground, in accordance with manufacturers instructions. Active external antennas were placed 6 meters above the ground, again following given instructions closely. Indoor active antennas were placed near a window on the 1st floor of the building. Screened antenna switches were used to allow each antenna to be connected to a Racal 117, ICOM IC-R70, JRC NRD 535 and a Kenwood R-5000. We tested reception of AM broadcast stations, SSB marine traffic, and standard telex. A professional RACAL preselector (bandwidth 100 kHz) was used to determine which interference products were produced by the antenna, and which were receiver generated. A HF spectrum analyzer was used to look at received signals, and a signal-strength meter was connected to a pen recorder for the analysis of fading. We listened between 20 kHz up to 30 MHz, though above 25 MHz signals were very weak due to propagation conditions.

We looked at ground-wave reception of 1008 and 747 kHz mediumwave, short and mediumwave transmitters between 500 and 1500 km distant, and distant signals which had travelled up to three hops over the ionosphere.

We listened during the day and during the night. Total listening time was in excess of 400 hours in the first five months of 1992. In presenting the results we used an adapted version of the SINPO code. A rating of 11111 is the WORST, 55555 is the BEST case.

S - Signal Strength

This is the average signal strength above atmospheric noise of a number of stations within given frequency range.

S Rating	Meaning	Signal Level	"S" meter reading
S5	Excellent	50 microvolts or more	S9+
S4	Good	12.5 - 50 microvolts	S7 - S9
S3	Fair	1.6 - 12.5 microvolts	S4 - S7
S2	Poor	0.4 - 1.6 microvolts	S2 - S4
S1	Barely usable	less than 0.4 microvolts	< S2

I - Intermodulation Interference Signals.

We looked for intermodulation products and these are rated as follows:

Rating	Meaning
I5	No intermodulation products detected
I4	Very few intermodulation products detected
I3	A few strong intermodulation products found
I2	Several strong intermodulation products found
I1	Large numbers of intermodulation products found

N = Noise

This is used to define the level of atmospheric and man-made noise picked up by the antenna while listening to weaker signals. We rated noise as follows:

Rating	Meaning
N5	barely audible
N4	slight noise
N3	fair noise
N2	high noise level (no DX possible)
N1	extremely high noise level (useless for DXing)

P = Propagation Disturbance.
This letter gives the degree to which the antenna was affected by fading effects. Small active antennas are clearly more susceptible to multi-path propagation.

Rating	Meaning
P5	Not affected.
P4	Slightly affected
P3	Fair level of fading
P2	Seriously affected by fading
P1	VERY seriously affected by fading.

O = Overall Rating of the antenna.
We judged the antennas on their overall electrical performance, bearing in mind the S, I, N, and P ratings as follows:

Rating	Meaning
O1	excellent
O2	good
O3	fair
O4	poor
O5	unusable

NOTES

We were careful to follow a strict listening pattern to the same stations throughout the tests. These were repeated at several times of the day to give average SINPO ratings. When checking the table, remember to check more than just the "O" column. Antennas show marked differences depending on the frequency and polarization of the received signal. In the 60 meter tropical band, the ARA-60 gives a much better signal than the DATONG AD-370. But the higher level of intermodulation products at 7 MHz leads to a poorer overall rating than the DATONG for short-range reception between 3.2 and 9 MHz. For long-range reception, the signal on the DATONG AD-370 drops back and fading increases, so that results are similar for both active antennas.

The results for each antenna are relative to others in the test. If a particular antenna is given a poor rating for a particular range, we do not mean to imply that DXing with this antenna is impossible. But performance is better (and therefore the chance of a readable signal) on other antennas. Indoor antennas always produce between 20 and 30 dB less signal than outdoor types. The "S" values are measured using the receiver at maximum sensitivity.

GENERAL WARNING

Never mount an antenna so that if it breaks it would fall onto a public highway or footpath. Never hang the antenna above telephone or high-voltage power lines. Use proper climbing equipment if you are working on the roof, tying one end of a nylon cord around your waist, and the other end around a firm anchor point (e.g. a chimney stack). In order to prevent birds accidentally flying into the antenna, a piece of aluminum tape or foil attached to certain portions of the antenna wire may help.

THE ANTENNA TEST: RESULTS

"ALPHA DELTA SLOPER"

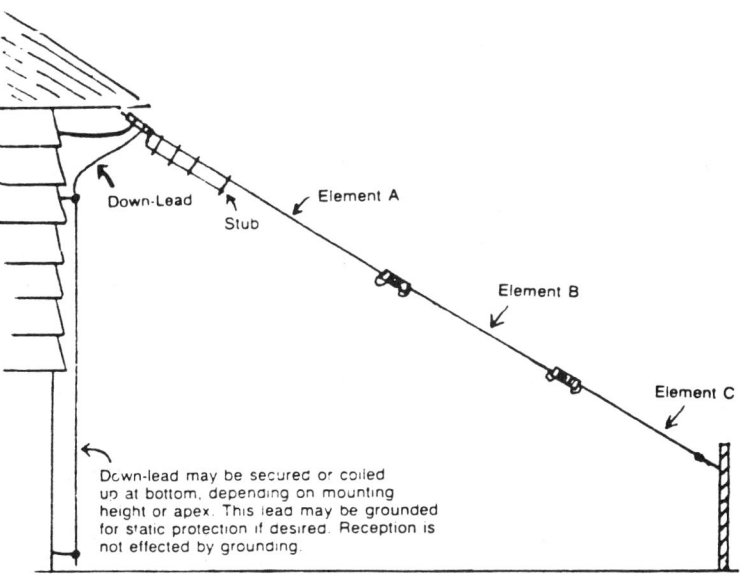

This passive wire antenna is very popular in North America, and is sold as an "all-band" aerial for the shortwave listener, covering 0.5 - 22 MHz. It consists of a wire around 19 meters long which is hung at an angle of 60

ANTENNA RESULTS: The higher the number, the better!

	A.D.Sloper	T2FD	DX Listener	Magnetic Balun	Dipole	Supr. Listener	DX-1	Liniplex	DX-7	ARA-60	Datong AD-370	Sony AN-1	Yaesu FRA-7700	Frequency Range
S	2	1	1	1	1	1	5	5	5	5	2	2	2	VLF
I	5	5	5	5	5	5	4	4	4	3	5	5	5	Time signals, beacons
N	4	5	5	5	5	5	3	4	3	3	4	5	5	20-200 kHz
P	5	5	5	5	5	5	5	5	5	5	5	5	5	
O	3	1	1	1	1	1	4	5	4	4	2	2	1	
S	2	1	3	3	1	1	5	5	5	5	3	2	3	LF
I	5	5	5	5	5	5	4	4	4	3	4	5	4	Broadcast signals
N	4	5	5	5	5	5	4	5	4	3	4	4	3	200 - 518 kHz
P	5	5	5	5	5	5	5	5	5	5	5	5	4	
O	3	1	3	3	1	1	3	4	3	3	3	2	3	
S	3	2	4	4	2	2	5	4	4	5	4	3	4	MW1
I	5	5	5	4	5	5	4	4	4	4	4	4	3	520-1602 kHz
N	4	5	5	5	4	4	4	4	4	3	4	4	3	local signals
P	5	5	5	5	5	5	5	4	5	5	5	5	5	
O	3	2	4	4	2	2	4	4	4	4	3	2	3	
S	2	1	4	4	2	1	5	5	4	5	3	1	3	MW2
I	5	5	5	4	5	5	4	4	4	3	4	4	3	520-1602 kHz
N	4	5	5	5	4	4	4	4	4	3	4	4	3	DX signals
P	4	4	4	4	4	3	4	4	4	3	4	4	5	>1500 km
O	2	1	4	3	2	1	4	4	4	3	3	1	2	
S	4	4	4	4	3	3	5	5	4	5	5	2	4	SW1
I	5	5	5	5	5	5	3	3	3	2	3	5	4	1.8-3.2 MHz
N	4	5	4	5	4	4	4	3	4	3	4	4	3	SSB, Morse, RTTY
P	4	3	4	3	4	3	4	4	4	3	4	3	5	up to 500 km
O	4	4	4	4	4	3	4	4	4	3	4	3	3	
S	3	3	3	3	2	2	5	4	4	5	3	1	3	SW1
I	5	5	5	5	5	5	3	3	3	2	3	5	4	1.8-3.2 MHz
N	4	5	4	5	4	4	4	4	4	3	4	4	3	SSB, Morse, RTTY
P	3	3	3	3	3	2	4	4	4	3	3	3	3	around 1500 km
O	3	3	3	3	3	2	4	4	4	3	3	2	2	

A.D.Sloper	T2FD	DX Listener	Magnetic Balun	Dipole	Supr. Listener	DX-1	Liniplex	DX-7	ARA-60	Datong AD-370	Sony AN-1	Yaesu FRA-7700	Frequency Range
S 5	4	4	4	3	3	5	4	4	5	4	2	3	SW2
I 5	5	5	5	5	5	5	4	4	4	4	4	4	3.2-9.0 MHz
N 4	5	5	5	4	5	5	5	5	3	4	4	3	Broadcast, Fax, SSB
P 4	4	4	4	4	3	4	4	4	4	4	4	4	distance 100-1500 km
O 5	5	5	5	4	4	4	4	4	3	4	3	3	
S 5	5	5	5	3	3	5	4	4	5	3	2	2	SW2
I 5	5	5	5	5	5	4	4	4	4	4	4	4	3.2-9.0 MHz
N 4	5	5	5	5	5	4	5	4	3	4	4	3	Broadcast, Fax, SSB
P 3	4	4	4	3	3	4	4	4	3	3	3	3	>1500 km
O 4	5	5	5	3	3	5	4	4	3	3	2	2	
S 5	5	5	5	5	3	5	5	5	5	5	3	3	SW3
I 5	5	5	5	5	5	5	5	5	4	4	5	4	9-15 MHz
N 4	5	5	5	5	5	4	4	4	3	4	4	3	Broadcast, Fax, SSB
P 4	4	4	4	4	3	5	5	4	3	3	4	4	<1500 km
O 5	5	5	5	5	4	5	5	5	5	5	4	3	
S 4	5	5	5	4	2	5	5	4	5	4	2	2	SW3
I 5	5	5	5	5	5	5	5	5	4	4	5	4	9-15 MHz
N 4	5	5	5	4	4	4	4	4	3	3	4	3	Broadcast, Fax, SSB
P 3	4	4	4	3	3	5	5	4	5	3	3	3	>1500 km
O 4	5	5	5	4	3	5	5	5	5	4	2	2	
S 4	4	4	4	3	3	5	5	4	4	4	3	3	SW4
I 5	5	5	5	5	5	5	5	5	4	4	4	4	15-30 MHz
N 4	5	5	5	4	4	4	4	4	3	4	3	3	Broadcast, Fax, SSB
P 4	4	4	4	4	3	4	4	4	4	4	3	3	<1500 km
O 4	5	5	5	4	3	5	5	5	4	4	3	3	
S 4	4	4	4	3	3	5	5	4	4	3	2	2	SW4
I 5	5	5	5	5	5	5	5	5	3	4	4	4	15-30 MHz
N 4	5	5	5	4	4	4	3	3	3	3	3	3	Broadcast, Fax, SSB
P 3	4	4	4	3	3	3	3	3	3	3	3	2	>1500 km
O 4	5	5	5	4	3	5	5	5	4	4	3	2	

degrees. The highest section of the antenna should be at least 8 meters above the ground. The design is based on the Zepp antenna. Connection to the receiver is via a 50 Ohm coaxial cable. The wire itself is divided into three sections, separated by low-Q RF choke resonators. Each element is designed to give optimum results at a different frequency. The cable connecting point is a UHF (SO239) connector, attached to a piece of aluminum. Although the wire gauge chosen is resilient against very rough weather, and the resonators are protected by paint which is resistant to UV light, the connector connection is open to the elements. Although low impedance, it would be a good idea to make the connection and then weatherproof the joint with suitable material (such a silicone rubber).

Performance wise the antenna gave fair to good results above 1.8 MHz. The noise-level of the Sloper is higher than a T2FD, but the bandwidth is larger. The antenna shows directional properties at higher frequencies, and gives better results for short-range signals than DX (low-elevation angle). At US$69.95 excluding shipping in the United States, this antenna is a good all-round performer. Shipping costs may make it rather expensive in other parts of the globe. Further information from Alpha Delta Communications Inc, P.O. Box 571, Centerville, Ohio 45459 USA. Tel +1 513 435 4772.

DIPOLE

As a reference in the WRTH tests we constructed a dipole using 2.5 mm thick wire, and a length of 12 meters. The tuned frequency for this turned out to be 12.5 MHz. The center connection to the 50 Ohm cable was via a 1:1 Balun transformer. This ensured that the lead-in cable does not affect the antenna diagram, nor affects the dipole's performance. It was mounted north-south at a height of 10 meters.

The dipole in our tests performed quite well between 9 and 15 MHz. But it gave variable results per station, depending on the polarization and direction of the received station. Bearing in mind that a T2FD is easier to construct, a dipole is only really an advantage if you want to study a particular part of the spectrum. When used around its resonant frequency the dipole is much less sensitive to man-made noise than the other two sloping antennas.

SUPREME LISTENER

A passive antenna for shortwave from New Zealand. It consists of two aluminum tubes of 19 and 22 mm which are screwed together. At one end is a metal block in which 5 radiators need to be screwed. The radials are made of glass fiber around which copper wire has been helically wound. Each radial has a different resonant frequency which lies in one of the major SW BC bands. The manufacturer claims a coverage of 5.9 - 22 MHz. Each radial is electrically connected to the aluminum pipe, so that the whole antenna is one large radiator itself. The height is 2.7 meters, the diameter 1.4 meters. Underneath the antenna is a screw to connect the coax cable. The inner braid is attached to the screw, whilst the outer braid should be connected to a supplied "earthing wire". The other end of this wire has to be connected to an electrode buried in the ground.

The antenna is mounted on a single aluminum clamp. A second clamp is supplied to attach the antenna to a suitable mast, which must be a maximum of just 28 mm thick. The impedance of the antenna is claimed to be 50 Ohms.

It should be noted that if a standard metal mast is used to mount the antenna on a roof, this also starts acting as part of the aerial, with unpredictable results. If the metal supporting mast should touch an earthing point, then the antenna is short-circuited. When mounted on the side of a house, moisture during wet weather on the brick wall was enough to reduce signal strength. The length of the earthing wire was also found to affect performance.

Mechanically we found this construction to be rather weak - fine for good weather, but easily blown away in the first storm. The stability of the radiator head left a lot of be desired. In our example it rocked about until taped down with water resistant tape. We followed the manufacturers' instructions and mounted it 6 meters above ground. The earth wire was dropped vertically to an excellent earth, and an isolated mount was used. Although quite a clever idea, this antenna does suffer quite strongly from fading effects. Since the results are very dependent on the quality of the earth connection, predicting how it will perform in a particular location is difficult.

Price of the "Supreme Listener" is NZ$136.20, excluding shipping. Further details from Tricity House Ltd, 209 Manchester Street, Christchurch, New Zealand.

THE T2FD ANTENNA

The "Terminated Tilted Folded Dipole" is widely used in professional applications. It was first described by C.L. Countryman in the November 1951 edition of the American amateur radio magazine "QST". Recent publicity by the World Radio TV Handbook and other publications has lead to a resurgence of interest amongst the hobby community too. The T2FD consists of a folded dipole terminated at one side with a resistor. It is sometime also referred to as a "squashed rhombic", although the performance when compared to a full-size rhombic is inferior. However, few people have the space for a proper rhombic antenna.

The impedance at the opposite ends of the T2FD antenna is around 500 Ohms. You could try to use open feeder or twinlead as the feeder to the receiver, but in practice any nearby metal objects (e.g. guttering, window frames etc.) will have such a detrimental effect that it is much better to match the antenna to 50 Ohm coaxial cable via a 10:1 balun transformer. The T2FD is one of the few broadband antennas for shortwave listening, offering a bandwidth of 6:1 for listening purposes. Our tests showed that a length of about 15 meters gives reception between 3.5 and 25 MHz, and a constant sensitivity over the entire range.

For the range between 3 and 30 MHz, this antenna gave the best overall results. The noise level is exceptionally low, and a T2FD is equally suited to short or long-range signals. Apart from a few dips, the antenna diagram shows an omnidirectional performance.

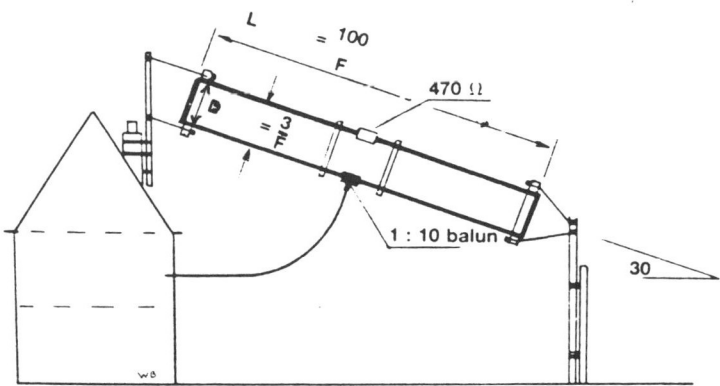

The diagram explains the antenna's appearance. The advantage of the design is that you can choose which section of the HF band you want it to cover by changing the dimensions. The formula is simple. The length of the antenna (in meters) is 100 divided by the desired lowest frequency in MHz. The distance between the two parallel wires is 3 divided by the frequency in MHz. The bandwidth of the antenna is about 1:5. This means that a 20 meter antenna will give coverage between 5 and 25 MHz. In practice it will work between 3 and 30 MHz, but you may find performance dropping off slightly at either end of the range. The distance between the parallel wires should then be 60 centimeters. By ensuring a constant impedance throughout the length of the antenna, (a T2FD is a "travelling-wave" antenna), the T2FD is also less prone to distortion due to multi-path fading.

Independent tests have shown that when compared to dipole or long-wire antennas, the background noise is not only much lower, but weak (DX) signals suffer from reduced distortion. Sometimes this can make the difference between being able to understand a weak station or not. In the case of data reception, lower distortion reduces the number of errors. If you hang the T2FD at an angle of 30 degrees with respect to the ground, then the antenna pattern shows so many lobes, you can regard the antenna as sensitive to signals from all directions (i.e. omnidirectional). Slight variation (between 20 and 40 degrees) is allowed but not outside these limits or you will notice an increase in man-made noise pick-up. Although a tuned dipole should give better gain than a T2FD, our experience indicates there is not much difference. Remember though that the T2FD outperforms the dipole when the receiver is tuned outside the limited frequency range of the tuned dipole.

The ends of a dipole, trap-dipole, and longwire antennas have a high impedance. This is a problem when the wire runs in the vicinity of conductors such as metal roofs, wet trees, and the like. The T2FD has less trouble because of its medium and constant impedance. The conductivity of the ground under the antenna has little influence on the performance. The height of the lower end of the T2FD does not have to be more than 1.5 - 2 meters above the ground surface.

If there are so many positive aspects to the T2FD, there must be a reason why the antenna is only gaining popularity again now. This is because the original T2FD fed into a symmetrical open feeder directly to the receiver. This is not only difficult to install, but it is tricky to effectively couple a 50 Ohm symmetrical feeder onto the 50 Ohm asymmetrical antenna input of most modern communication receivers. This problem has now been solved.

INTERFERENCE FIELD

There is an interference field around every house! It is caused by electrical equipment inside the dwelling, especially devices such as dimmers, TVs, computers, video-recorders, and thermostats. Even if you don't use such apparatus when listening, there is a good chance that the neighbors do. The mains electricity cables also bring their interference into your house, and this in turn is radiated by all the electricity cables. This interference field stretches some 5 meters around the house, and about 1 meter above it. Even though the T2FD is less sensitive to man-made interference than open dipole and longwire antennas, it is always a good idea to hang the T2FD away from this interference field. In many cases this is possible by attaching the lower end of the antenna to a pipe fixed to the top of a garden shed, or using a metal pipe that has been firmly anchored into the ground. If the nylon cord on the high end of the antenna does not reach the support, then use a longer piece. You can obtain 3 or 4 mm nylon cord through most hardware stores.

EARTHING

During thunderstorms, but even during dry and wet weather, static can build up on the antenna. This must be allowed to flow to earth! Getting rid of the static not only protects the sensitive input circuitry of your receiver, but also reduces the background noise. At any one moment, some 30,000 thunderbolts are around somewhere on the planet (especially in the tropics). All this leads to unwanted natural "noise". The RF Systems T2FD (mentioned below) is special in this case because the special matching transformer offers a galvanic path to earth, PROVIDING you use a proper earth connection.

Ideally the earth connection consists of a copper wire which is a minimum of 2.5 mm thick. Make sure this makes a perfect electrical connection with a metal spike banged into the ground. You can also connect it to a lightning arrester if available. If you can't find a spike, use a zinc bucket or chicken wire (about 5 square meters in all) buried a few inches under the surface of the soil.

You can connect the earth in two ways. The best place to connect the earth is at the point where the coaxial lead-in enters the house. Carefully peel back the black protective coating on the cable, making certain not to damage the copper braid underneath. Use solder to make a good earth connection with the outer braid of the coax, taking care not to melt the insulation. Make the earth connection completely water-tight using self-vulcanizing tape or silicone rubber. It is very important that this connection doesn't get damp, or the performance of the antenna at higher frequencies will suffer considerably.

The second way of connecting an earth to the system is on the rear side of the receiver. Don't rely on the "earth" connection that is part of the house electrical system. It is contaminated with noise from all the other appliances in the house. The earth is often overlooked by many listeners, despite heavy investment in buying a good receiver and antenna.

Note that a good earth is no protection against a direct hit by lightning. If a thunderstorm is in the direct vicinity, disconnect the antenna from the receiver. Put the antenna plug into a plastic bag to protect it from moisture, and hang it out of the window.

DO-IT-YOURSELF T2FD CONSTRUCTION

Standard 50 Ohm coaxial cable should be easy to find. Note that although the length is not important, you should not allow the lead-in coax to run parallel with the antenna wire in close proximity. Let the coax drop a few meters away from the antenna before turning towards the receiver.

We used 5/8 inch plastic piping to spread the antenna wires apart. Do-It-Yourself stores have this type of piping in plentiful supply. On the outside two plastic supports three holes are bored at either end (see diagram). The inner two are used to support the antenna wire so that it cannot shift. The outer hole is used for the insulating nylon chord (3 mm thickness is sufficient).

The wire you use for the antenna should ideally be made of pure copper and between 3 and 5 mm thick. Do not underestimate the strain this antenna will be under. If you choose wire that is too thin your antenna will self destruct within days! If proper antenna is not available, three core household cable is also a solution. Wind the three cords together at the ends so that each core is used in parallel. The disadvantage with this cable is that it is liable to stretch. After a few weeks the antenna may start to sag. If this is the case, tighten the nylon supporting cord. Usually this is enough to solve the problem.

THE MATCHING BALUN

The T2FD we are describing here has an impedance of 500 Ohms over its entire length. Because that is quite high, there is little influence on performance from trees or roofing. Insulation should provide no problems. But because 50 Ohm antenna cable is used for the downlead, the 10:1 balun transformer is needed. You could construct the balun yourself, but it is not something for the beginner. In the middle of the upper antenna wire is a resistor. For receive only purposes, a 470 Ohm 1 Watt carbon-type resistor is ideal. Don't use a wire-wound type antenna or performance will be affected.

Don't just break the antenna wire in two and solder the resistor in between. The first breath of wind and the resistor will snap in two. Take a piece of plastic piping. Drill in two holes and put in two metal bolts. In Europe, the size M4 is about right. Solder the wires to either side, and the resistor between the bolts. Then fill the pipe with some protective material (e.g. candle wax) to make it weather-proof. DO NOT use two 1000 Ohm resistors in parallel to obtain exactly 500 Ohms. The antenna works better when this terminating resistor is slightly below 500 Ohms.

Guy Atkins, writing in "Proceedings 1990", dealt with the T2FD in some detail too. He points out that if the 10:1 balun is sometimes difficult to find at electronic stores, similar results are possible with a 4:1 balun, a 390 Ohm resistor (a carbon NOT wirewound type), and a 75 Ohm coaxial feeder cable (RG-59 or RG-6). The transformer is mounted exactly in the middle of the lower antenna wire. If you want to experiment further, we can highly recommend the article by Guy Atkins. Proceedings 90 is designed for the advanced listener, containing over 200 pages of fascinating and practical advice. Details from Fine Tuning Publications, RRT #5, Box 14, Stillwater OK 74074 USA Price: US$19.50, (shipping costs on request)

READY MADE T2FD

The Dutch company of RF Systems has experimented with the T2FD for several years, continually improving on the design. By analyzing the problems from different angles, and trying various materials, the good points of the original design were improved upon. The new design means that common coaxial cable (RG 58/u) can be used as a lead-in to the receiver, eliminating interference from equipment such as computers, dimmers and fluorescent lights.

To start with, the characteristic impedance of the RF Systems T2FD has been raised to 550 Ohms. Thanks to the development of a special frequency-compensating 11:1 transformer, the antenna is perfectly matched to the 50 Ohm coaxial cable and the input impedance of most communications receivers. By using frequency compensating technology, the bandwidth of the antenna (at -3 dB points) works out to be 3 - 35 MHz even though the length is just 15 meters. This special transformer not only ensures a perfect symmetry in the antenna across the frequency range, but also that the coaxial cable is isolated (from the high-frequency point of view) from the antenna. In that way, interference signals that are picked up by the outer braid of the coaxial cable do not interfere with signals picked up by the antenna, nor do they influence the reception pattern.

In addition the transformer ensures that the antenna-wire on the T2FD has a galvanic connection to earth, so that any static build-up that may occur, due to nearby thunderstorms, is safely discharged to earth. This not only protects the sensitive input circuitry of the receiver, it reduces the atmospheric noise which is generated as a result, especially in the region between 3 and 7 MHz.

All sensitive components are in the center piece, which has been filled with polyurethane foam. The antenna is therefore completely waterproof. The connection between the antenna and coaxial cable is sealed off with a rubber sleeve which slides over the PL 259 connector. The antenna is made of UV-resistant plastics and stainless steel fasteners. The antenna is delivered complete, and only needs to be hung. But the coaxial cable between the antenna and receiver is not included with the kit. The RF Systems T2FD antenna is supplied through the world-wide agent for RF Systems products: Doeven Elektronika, Schutstraat 58, 7901 EE Hoogeveen, The Netherlands.

RF SYSTEMS "MAGNETIC LONGWIRE BALUN"

Longwire antennas are still some of the best antennas for shortwave listening, providing they are of reasonable length and height. Unlike active antennas, they can't suffer from intermodulation products, and in general they are less sensitive to fading due to multi-path propagation.

But there are drawbacks. The connection between the antenna and receiver is a single wire, which must be isolated from the antenna supports. The lead in wire is also extremely sensitive to man-made interference, especially computers and fluorescent lights. And during thunderstorms static charges, can build up on the antenna wire which can damage sensitive receiver input circuits.

At first sight it might seem a good idea to use a screened cable between the longwire and the receiver. But there is an impedance mismatch between the coaxial cable, and the longwire antenna.

RF systems have come up with a compact, weatherproof matching unit. It consists of a special patented transformer with magnetic field coupling. The degree of coupling changes with frequency. The end result is a simple device which you connect to the end of a longwire antenna (or in the middle of a "T" antenna). The other end of the unit has an SO-239 socket designed to match 50 Ohm coaxial feeder cable. The advantages are immediate. The lead-in cable is immune from local man-made interference, and doesn't have to be isolated from its supports. The longwire has a galvanic connection with earth, thus avoiding problems due to static build-up. The antenna circuit is electrically isolated from the receiver - the transfer of signal takes place magnetically. This leads to a reduction in overall background noise. The balun transformer inside the unit ensures a high resistance load on the antenna, irrespective of the frequency being tuned.

The "Magnetic Longwire Balun" from RF Systems performs best using 12 meters of wire. Longer is possible, but if you use a shorter length then performance on medium and lower shortwave bands drops off significantly. The longwire needs to be isolated from its supports.

LONGWIRE COMPARISON

The Magnetic Longwire Balun was compared with a standard longwire of identical length. On longwave, the signal strength on the Balun antenna was lower , but the atmospheric noise was also lower. Likewise the man-made interference level (from inside the house) was less noticeable on the balun longwire.

If you have the space, 99 Dutch Guilders (£28) is not much to pay for a simple, yet effective magnetic longwire matching unit. There are baluns on the market at a cheaper price, but they are not weatherproofed. The price mentioned does not include shipping. Originally launched back in November 1990, this product has a lot of potential for the experimenter. It comes with an English language fact sheet with tips on constructing a good longwire antenna. Details from Doeven Electronics (address given above). EEB and Universal Radio carry the MLB for the North American market.

DRESSLER ARA-60

The German company of Dressler have built up a fine worldwide reputation for active antennas, both shortwave and VHF. Our research shows they give honest advice to customers, explaining clearly situations where an active antenna is a suitable choice. In the 80's, Dressler built-up an excellent world-wide reputation for their ARA-30 active antenna. This has now been replaced by the ARA-60, a much improved antenna in many respects.

The new ARA-60 consists of a white plastic pipe. The bottom of the pipe is blocked with a PVC disk. From the outside you see the S0239 connector socket. On the inside is a printed circuit board, plus two vertical wires which run up the sides of the pipe, and are held tight at the top of the antenna by a black plastic cover. The wires are prevented from rattling inside the pipe by a piece of foam plastic. The ARA-60 is designed to be mounted vertically with the antenna cable dropping down away from it. The screened coaxial cable can then be fed indoors to a control unit. On the side of the power unit is provision to connect one receiver. The power unit works off AC mains or a 12 volt DC supply (e.g. a car battery).

Measurements show that below around 200 kHz, the performance of the antenna drops off markedly. So don't use the antenna for VLF work. A resonance peak around 157 MHz is generated from within the antenna because of the transformers used in the amplifier. It is curious that the ARA-60 has the highest level of amplification in the mediumwave region. Not many modern receivers can cope with strong MF signals on the receiver input and perform well on the 60 & 90 meter tropical bands. It is possible to regulate the output of the antenna (using a screwdriver), but when amplification is needed (e.g. above 20 MHz) you'd have to turn it up again. This attenuator can be set to reduce the output voltage of the antenna by 25 dB.

The most important measurement on an active antenna is the level of intermodulation. When two strong signals are received at the front end of the amplifier inside the antenna, "ghost" signals may be generated further up or down the dial. These may block the reception of real signals on those frequencies. The ARA-60 gave a measurement of +32 dBm. This is a excellent figure, though not quite as good as the +40 dBm we obtained from the DX-One active antenna from RF Systems. However, the ARA-60 is far more compact than the DX-One, and it's around US$50 cheaper. We recommend that the ARA-60 is powered from the mains. When we measured performance using a 12 volt DC power supply, the intercept point dropped to +27 dBm. That is still usable, but you're not getting the best out of the product.

The power unit supplied with the antenna delivers 15.2 volts DC when powered at 220 volts. That's quite high since Dressler has decided to drive the antenna amplifier to its maximum. That means the housing of the IC amplifier and the regulator gets quite hot, around 50 Celsius when the outside temperature was 20 Celsius. We think that in countries where the outside temperature is around 30˚C, this may cause the IC to overheat. The ARA-60 is not fitted with any mechanism to cope with static build-up during thunderstorms. The top of the antenna is water-proof, but moisture can get in underneath through the hole in the SO239 socket. It is therefore important to weatherproof the connector.

The ARA-60 performed well in our listening tests, especially on the tropical bands. Because it is vertically polarized, slightly higher levels of man-made interference were noted when compared to the passive antenna. If you use this antenna in an apartment, keep in well away from fluorescent lights!

At a price in Europe of around US$275 (US$200 through Gilfer Shortwave in New Jersey USA), the ARA-60 offers good value, and is a clear improvement over the ARA-30. More details from: Dressler GmbH, Werther Strasse 14-16, D-5190 Stolberg-Mausbach, Germany. Tel: +49 2402 71091. Fax: +49 2402 71095.

DX-1

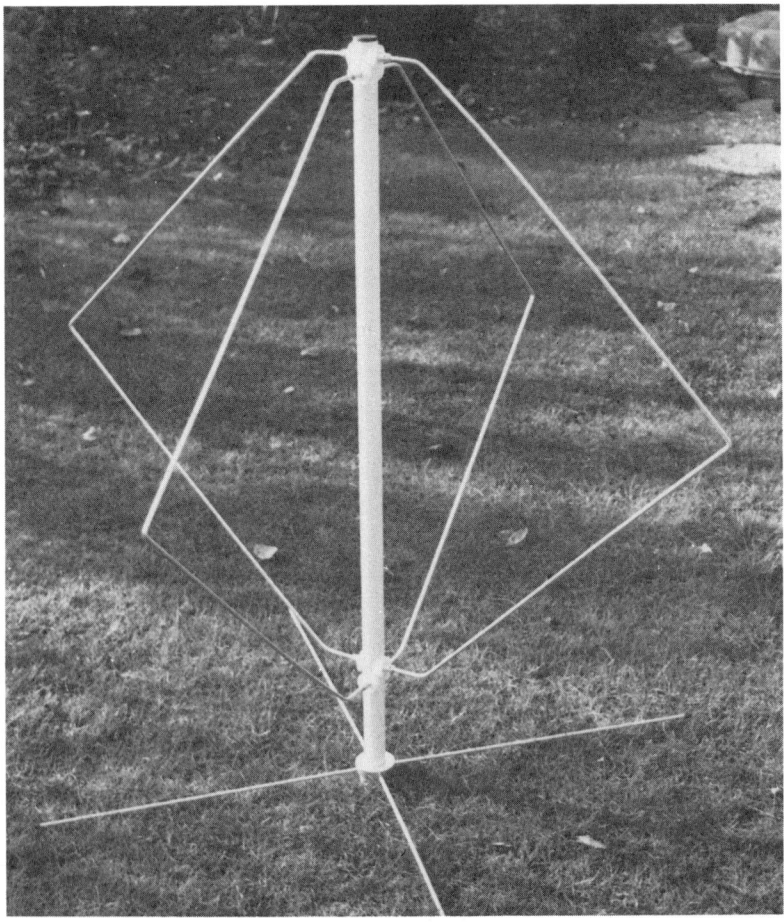

The DX-1 from RF Systems Inc is an active antenna, originally designed for military use. It claims a constant impedance of 50 Ohms between 50 kHz and 50 MHz (with reduced specifications up to 160 MHz), and is resistant to multi-path fading and polarization changes. The specifications state that it can be safely used in areas of high signal strength (e.g. on board ship). The consumer version is electrically identical to the military version, but the radials are made of aluminum, a simpler mount is offered, and a different indoor unit.

The DX-1 consists of a fiberglass pipe, in which a specially formed vertical element is mounted. Two single wire loops are attached to this pipe. Four earthing radials are connected underneath. The loops and the vertical rod are connected to the amplifier via broadband couplers. The amplifier is mounted inside the pipe, which is turn is filled full of polyurethane foam to weatherproof it. The antenna is 1.2 meters high, 1 meter in diameter.

The DX-1 is omnidirectional and offers a gain of 6 dB. If you use such an antenna near a transmitter site (include a radio amateur), the output voltage can rise very high: up to 15 volt peak to peak at 50 Ohms. The indoor unit (16 x 15 x 6 cm) is unique. It contains a 30 volt 500 mA power supply, and a gain control to control the amplification. Steps of +6, 0, -10, -20, -30, and -40 dB are offered. Two independent connectors are available, so the unit can feed two receivers at the same time. Finally, the control unit has a medium-wave attenuator which suppresses 540 - 1604 kHz by 30 dB when activated. This is designed for users living near strong mediumwave AM transmitters.

The DX-1 gave the best overall results of the active antennas tested. Whilst the DATONG AD-370 and DRESSLER ARA-60 gave better signals in some cases (dependent on polarization received), results from the DX-1 were consistently good. This was also the most expensive antenna we tested. The high intercept point (48 dBm) and resistance to multi-path fading were encouraging. Intermodulation problems were at a low level, and between 16 and 100 kHz (VLF) this was the only usable antenna we tested. Although outside the scope of this test, the DX-1 can be used for stand-by use on the VHF FM broadcast band, aircraft and 144 MHz 2 meter ham band. The noise levels at these high frequencies though is rather high.

For most consumer communications receivers, the use of the variable attenuator will help to avoid overloading problems. This active antenna though cannot be mounted on a window ledge. It needs a flat roof or a chimney for support. The price in The Netherlands is 699 Dutch Guilders (about US$ 300 or UK£180), excluding carriage.

DX-LISTENER

The DX-Listener is a 15 meter long transmission line antenna for the frequency range 100 kHz to 35 MHz. The antenna comes complete with a simple-to-operate indoor control unit. This allows the user to switch the antenna so that it operates either as an open or as a terminated transmission line. Note however that this does not imply the antenna can be used for transmitting (such as for amateur radio work). It is strictly designed for reception only.

In the "Wide-Band" position, the DX-Listener works as a high efficiency antenna with a frequency range between 100 kHz and 25 MHz. In this mode the antenna gives remarkable results especially on medium & longwave, as well as very low frequencies. When the indoor control unit is switched to "Low Noise", the transmission line construction switches from open to terminated. This gives a much lower noise level between 3 and 35 MHz, due to reduced sensitivity to static and nearby man-made noise. Signals received in this low-noise position also suffer less from the distortion introduced by multi-path propagation. In this respect the performance of the DX-Listener is comparable to the T2FD design. The DX-Listener therefore combines the best of both worlds, giving the serious listener a choice between maximum signal strength or minimum noise and fading.

The supplied indoor unit has a screened metal case. Inside is a power supply. Yet the DX-Listener is not an active antenna. The indoor unit is used to control RF relays which are sealed inside the center part of the DX-Listener. These relays have gold contacts enclosed in a vacuum, which switch between matching and routing networks built-inside the antenna. The antenna circuit itself is therefore completely passive, thus unwanted intermodulation products, produced by electronic components, cannot arise.

The indoor unit power supply is designed to work off 100-240 volts AC (50/60 Hz). The transformer is double isolated and uses so little energy that the indoor unit can be safely permanently connected to the household current supply. A connecting cable between the indoor unit and the receiver is included, plus a PL 259 plug. The DX-Listener comes completely assembled and is ready to hang.

We hung the antenna according to the instructions, and also in other positions to see just how critical the mounting really is. In order to get the best results this antenna needs to be hung at an angle of 30 degrees, so the top of the antenna needs to be around 7.5 meters higher than the bottom. The antenna comes complete, ready to hang, and has a very sturdy construction which is completely water and rust proof. The center spacer contains the connector for a PL 259 plug. The plug and connector are made waterproof by using a simple and effective mechanism: covering the plug with a silicone rubber teat used for a baby's bottle, and feeding the coaxial cable through the nipple. It's important that the coaxial cable drops away vertically from the antenna for about 2 meters before bending in any direction that's required.

Once mounted, the DX-Listener is two antennas in one. When the indoor control unit is switched to the "wideband" position (0.05 - 40 MHz) its performance is very similar to a longwire connected to the receiver via a magnetic balun. Switching to the low-noise position transforms the antenna into a T2FD (3 -30 MHz) construction. If you listen to low-angle DX signals, switching to the low-noise position often produces dramatic results...the atmospheric and man-made noise just drops away. The output of the T2FD is not as high as in the wideband position, but it is a much quieter antenna. That said, it is useful to have both positions so as to be able to choose the antenna which gives the best performance at that moment. Sometimes the theory doesn't match the practice.

The DX-Listener antenna performs consistently throughout the shortwave spectrum. But you must follow the mounting suggestions. We tried hanging the antenna horizontally between two chimneys, but the drop in noise-level in the T2FD position vanished. In one test location we disconnected the earth connection from the receiver, and noticed a rise in background noise. Perhaps the best advantage of this antenna is the total absence of intermodulation products, found on even the best active antennas. At a price of US$350 in Europe, the DX-Listener is not cheap. A simplified and thus cheaper version is also available with just the T2FD performance, i.e. not suitable for reception below 3 MHz. But you can expect the DX-Listener to withstand severe storms and give noticeably better results. More details from: Doeven Elektronics, Schutstraat 58, 7901 EE Hoogeveen, The Netherlands. Tel: +31 5280 69679. Fax: +31 5280 72221. In the US, Universal SW in Reynoldsburg, Ohio and EEB, Vienna Virginia, carry RF Systems products. Lowe Electronics distributes them in the UK.

SONY AN-1

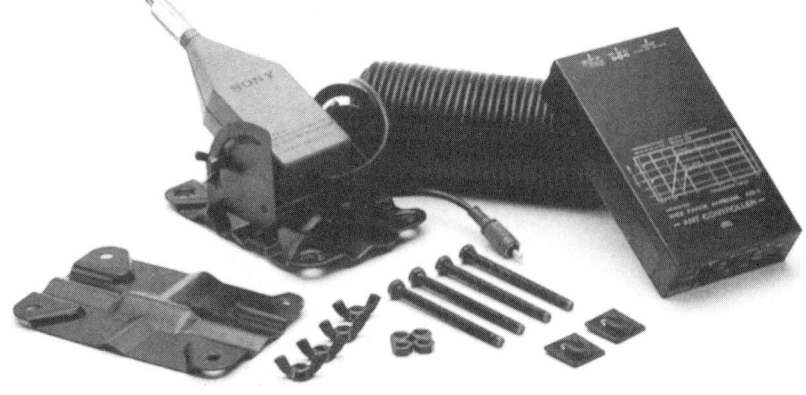

The SONY AN-1 consists of various sections, well-presented in special box with clear instructions. It consists of the antenna unit, an antenna controller (for power supply and connection box), a telescopic whip coupler, and ferrite antenna coupler. The active antenna consists of a weather-proof plastic box with an extendible telescopic whip. The gauge is much thicker than found on a portable radio, and its length is 1.5 meters. The supplied clamps allow the antenna to be mounted in various ways, including out on a window ledge. Permanently attached to the antenna unit is 12 meters of lead-in wire which plugs into the indoor controller box. This contains 6 AA penlight cells which SONY claim will last around 90 hours. This is fine for caravans, but if you mount the unit permanently you might want to try a 9 volt DC transformer unit. If use of the household supply results in much higher noise levels, rechargeable penlight cells may offer another solution. The controller also offers a 0/-20 dB switchable attenuator, plus a medium-wave 6 dB attenuator to reduce the effects of nearby AM transmitters.

The AN-1 uses a single Field Effect Transistor (FET) as the active element, and SONY gives the range as 150 - 30,000 kHz. The telescopic whip coupler is designed for use with a portable with no provision for an external antenna. For better medium/long-wave reception, the ferrite coupler is recommended for use with portables. We used a communications receiver with direct connections for the WRTH evaluations.

If you live in a concrete apartment block which completely shields out SW signals, then mounting a rather unobtrusive AN-1 just outside may help a lot. The active antenna comes with a number of different mountings, though some may not be strong enough if the site chosen is exposed to heavy gales. Our tests show that the efficiency of the AN-1 is rather low. Not much of the signal picked up by the telescopic whip is delivered to the FET amplifier. This means that intermodulation products are low, but weak distant signals are buried in the noise generated by the antenna itself. Although suitable for listening to stronger broadcasters, this is certainly not a DX antenna. If you try the AN-1 with a simple portable (such as the ICF-7600DS) in a building where signals do penetrate, then in most cases the receiver gets too much signal, and an active antenna is simply a noise box.

The Sony AN-1 is available through most Sony dealers. Recent prices in North America average around US$84.95 , and NZ$225 in New Zealand.

DATONG AD270/AD370

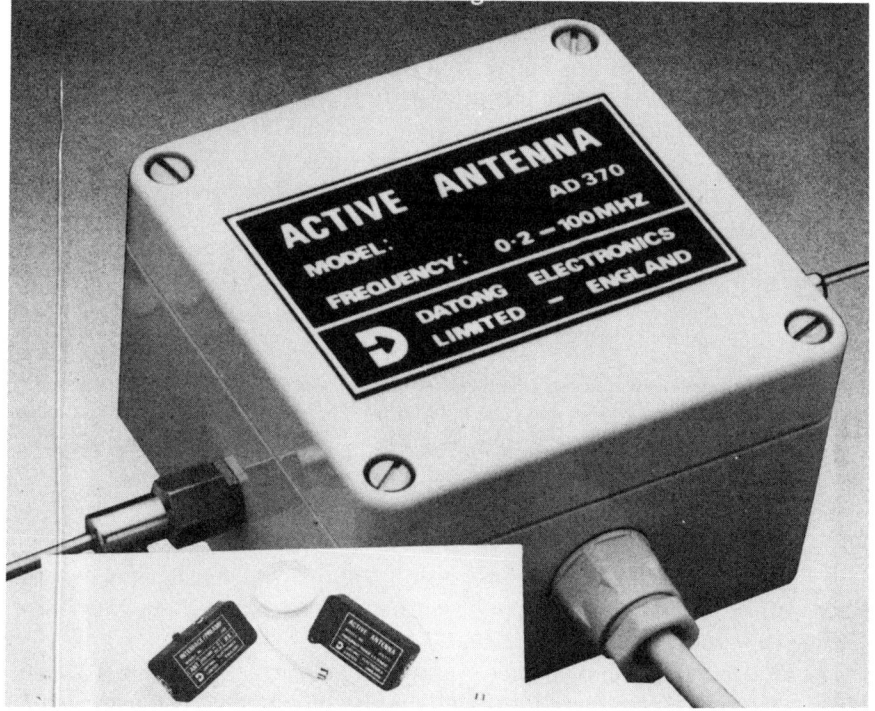

DATONG AD370

This has been a very popular active antenna, especially in Europe, although DATONG no longer seem to be actively developing new products for the radio hobbyist. DATONG still sell two models: the AD270 is designed for use indoors (e.g. in an attic) and the AD370 is weatherproofed for use outside. One sample of the AD370 unit purchased and mounted outside 8 years ago has stood up well to battering by North Sea gales. In the center of a mini horizontal dipole is the impedance matching unit. The electronics inside the plastic housing matches the impedance of the antenna to the 50 Ohm asymmetric coaxial cable. The lead-in cable is brought indoors and connected to a controller box near the receiver. A supplied 12 volt DC transformer unit plugs into the side of the controller, power being delivered to the antenna via the same coaxial cable. A 20 dB preamplifier can be switched in and out of circuit on the controller. We switched it in during our tests.

DATONG quote the range as being 200 kHz - 100 MHz, although sensitivity drops off above 30,000 kHz. Above 10,000 kHz the antenna experiences directional properties, but not below. Although the horizontal elements are less prone to nearby man-made interference, good positioning away from noise sources is recommended. We found the AD-370 noticeably quieter than the ARA-30, and slightly less noisy than the DX-1. Signal levels are less from the DATONG, but that had no effect on readability of signals. The low noise and acceptable number of intermodulation products makes this antenna score better than the ARA-30 on some bands. The lower efficiency of the antenna compared to the ARA-30 and DX-1, and the horizontal polarization meant that low-angle signals (3 or more hops) were noticeably weaker, sometimes dropping below the noise. The situation was reversed for signals from stations within Europe (less than 3 hops). The intercept point of the antenna was measured to be 27 dBm.

The price of the AD370 in North America is around US$149.95 through dealers. In Europe it retails for around £69.00.

PHASE TRACK LINIPLEX LOOP ANTENNA

This company has served the professional shortwave market for several years with a specialist receiver for re-broadcast purposes. Now they have released a loop antenna - for a price.

The loop arrived folded and takes around a hour to assemble. It is a vertically polarised antenna, and ideally should be mounted on a mast capable of rotation. The coverage is 50 kHz - 30 MHz, and especially for long and mediumwave, the ability to rotate the loop is desirable. Deep broadside nulls of around 27 dB were measured at the test site. It should be pointed out thought that the influence of the ground conductivity plays an important role here too. This will of course vary from location to location.

The loop's low-noise amplifier is also mounted on the outside mast, from which a coaxial cable feeder is led indoors to the power supply unit and then the receiver. The Liniplex instruction booklet is complete to the last detail in giving the best advice on how to set all this up. If the leads between the loop and receiver are very long, a second coaxial line can be connected in a symmetrical configuration to try and reduce the noise pickup. In our test case this wasn't necessary.

The antenna ideally needs to be position on a short mast just 1.5 meters above the ground. At this sort of height, weak DX signals coming in at a low angle will be screened if there are too many metal objects or tall buildings near the loop. Our listening measurements show that the loop is clear-

ly less sensitive to nearby man-made interference than most active antennas (such as the Datong AD-270). Still at the test site we found it was best positioned at least 5 meters away from the house. That said, since the loop is only 1 meter in diameter, it is very compact and unobtrusive. A switchable high-pass filter is included in the indoor power supply box to suppress the level of signals below 1.6 MHz. The small switch, however, is inside the plastic box, clearly designed to be used occasionally, not all the time.

The specifications state that the loop antenna will withstand winds of up to 100 miles per hour. However, the very thin wire used for what is in fact an array of small loops will certainly have problems in a heavy snowstorm. In our opinion the part of the Liniplex Loop needs to be far sturdier. Beware too of children....if they can reach the wires one "twang" will snap them.

The Liniplex Loop Antenna is not cheap. At £396.75 (including VAT but not shipping) this is the price of a budget communications receiver. Electrically it's performance was good, with intermodulation levels measured at +32 dBm at the test location....an excellent figure. Mechanically we feel the thin loop wire makes it is rather weak...especially as this is intended for semi-professional applications. Further information from Phase Track Ltd, 16 Britten Road, Reading, RG2 0AU, England. In Germany the unit is carried by Alltronic GmbH, Eichborndamm 178, D-1000 Berlin 51.

LINIPLEX / DX-1: A COMPARISON

The DX-1 from RF Systems is around half the price of the Liniplex Loop. We set up each of the antennas according to the manufacturer's instructions, and via a switching unit we were able to make direct comparisons. It turned out that each antenna displayed its own advantages. We leave it up to you to determine which characteristics you need, and how much that will cost.

The directional characteristics of the Liniplex were an advantage on medium and longwave - but we needed a mechanical rotator to physically turn the loop. The DX-1 is omnidirectional, and also sensitive to signals that are both vertically and horizontally polarized.

The Liniplex is vertically polarized. Above 5 MHz we noted some signals were slightly stronger on the DX-1 due to horizontal polarization. Some multi-path fading was less noticeable too on the DX-1. Signals from European stations were generally stronger on the Liniplex than the DX-1, especially above 7 MHz.

Both antennas have good specifications when it comes to intermodulation and cross modulation. Despite the proximity of high power transmitters, no spurious signals were noted. But then in an antenna of this price range, this is to be expected. The background noise on the 60 meter tropical band was lower on the Liniplex than the DX-1.

In terms of weather-proofing the sturdy construction of the DX-1 was superior to the Liniplex. The DX-1 should, ideally, be mounted higher than the Liniplex (4 meters above the ground as opposed to 1.5 meters). But in order to get weak DX signals the siting of the Liniplex is more critical, a clear horizon being preferred.

DX-7

Launched in August 1992, the DX-7 is a budget version of the DX-1 active antenna. It has slightly reduced performance characteristics, but it is also half the price of the DX-1.

The antenna is also far more compact, ideal for someone with limited space for an external antenna. Since the antenna is active it should be mounted on a metal pole outside the interference field of the house (at least 5 meters away from the building). The "coil" design is unique, enabling 2.3 meters of antenna to be wrapped round a sealed polyurethane pipe 50 cm long. The unit requires an indoor power supply. The DX-7 comes with a standard connection for a car battery (for portable use). An RF isolated power supply is also available as an optional extra, this operating on any AC power supply between 100-250 volts. Power is fed to the antenna through the coaxial cable (not supplied) and this may be anything up to 50 meters in length. The electronics for the antenna are housed inside the tube which is fully sealed from the outside elements. In order to avoid static build-up during thunderstorms we mounted the DX-7 on a grounded metal pipe (up to 40 mm in diameter is possible), threading the coaxial through the pipe.

The DX-7 produces noticeably less output than the DX-1 or the ARA-60. However, in comparison to the ARA-60, the intermodulation level is also lower - it is a medium efficiency antenna. The intercept point of our example was +24 dBm, lower than the ARA-60 and DX-1. It was a quieter antenna than the ARA-60 and more compact. The antenna costs f 399 in Holland, half the price of the DX-1. Further details from Doeven Electronics (address above).

CONCLUSIONS

If you have the space for a passive antenna (i.e. the T2FD, DX-Listener or Alpha-Delta Sloper) then results on shortwave (3 - 30 MHz) are MUCH better than on any active antenna we tested. Consumer-level active antennas produce unwanted intermodulation products, though the better brands did a better job of keeping this to a minimum. Problems between 1.6 - 3 MHz (mediumwave products) and 14 MHz (7 MHz harmonics) were common, as were 3rd order products.

The second, and often neglected problem was selective fading. Modulation from broadcast transmitters didn't sound as "clean" as with the passive antennas. The level of distortion was dependent on the depth of fading experienced. This was far more critical for utility listening. Many active antennas deliver too much signal to the front end of the receiver. That said, if you are considering one of these antennas, the top four tested were clearly the DX-1, ARA-60, DX-7 and AD-370. Refer back to the individual advantages and disadvantages, since the choice is dependent on what you are planning to listen to.

From 10 - 3000 kHz (50 - 1800 kHz in particular), active antennas easily beat the passive variety. Intermodulation products may cause some problems though, especially when listening to weaker signals. Manufacturers of active antennas often warn you not to place their device too high above the ground. They mean it. For interest, we put a DRESSLER ARA-60 on a 20 meter metal telescopic mast. The intermodulation level rose so high as to make reception of the shortwave spectrum unusable. Dropping from the standard 6 to 2 meters above the ground reduced intermodulation products considerably, but the weak DX signals were then buried in man-made noise. If you do pick an active antenna, then experiments with the height of the unit above the ground are often worthwhile.

INDOOR ANTENNAS

An indoor antenna for shortwave use is really a last resort. Higher susceptibility to local interferences (e.g. TV's) and screening effects of reinforced concrete walls are some of the reasons that indoor antennas run into problems. But what if such an antenna is the only solution?

YAESU FRA-7700

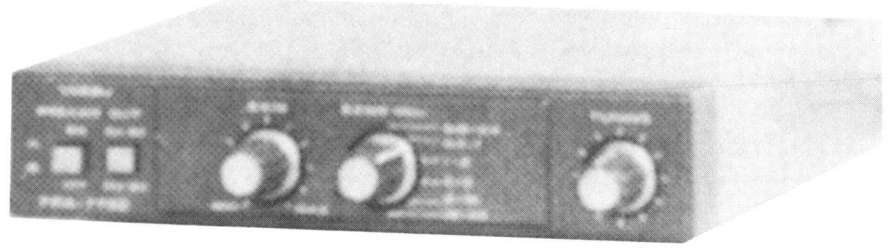

The Yaesu FRA-7700 was designed for use with the now discontinued FRG-7700 communications receiver, and the FRT-7700 antenna tuning unit. It is still being sold, as connection with other receivers (including the FRG-8800) is possible. If a 12 volt DC power supply is used, it is up to the user to solder a suitable 5 pin DIN connector to fit into the FRA-7700 unit. The supplied circuit diagram makes this quite straightforward. The FRA-7700 is different from other active antennas tested in that it is coupled to a LC network, i.e. a variable capacitor and a loading coil. A switch is used to select 6 sections of the band, depending on which frequency in the range 150 - 30,000 kHz is selected. This set-up increases the front-end selectivity somewhat (though not a great deal), resulting in some reduction of intermodulation products. We found quite a few 3rd order products in our tests though. The input stage is followed by an adjustable amplifier with 4 active elements, i.e. 2 FETs and 2 transistors. Gain can be adjusted between 0 and about +16 dB.

The FRA-7700 is well built in a metal box 17 x 17 x 3.5 cm. It can be put on top of the receiver, but it is VERY susceptible to man-made noise sources, e.g. nearby quartz clocks, dimmers, motors, etc. Although not much use for serious DX work, the FRA-7700 was better than a standard telescopic whip on a portable radio. The ability to tune it gave a sharp reduction in 2nd order intermodulation products. It was the only indoor antenna with a coverage from 150 - 30,000 kHz.

The price of the FRA-7700 is around ú57 in Europe, NZ$176 in New Zealand. It hasn't been advertised in North America for some years.

INTERCEPTOR LOOP ANTENNA

A indoor loop antenna can offer two main advantages. It is much less sensitive to atmospheric noise than a standard dipole or vertical antenna. It is also highly directional, which is why any loop must be rotatable, allowing you to "null" out interfering stations on the same frequency. Loops have the disadvantage that they don't deliver much energy, and with any amplification there is the danger of intermodulation problems. The A-428 from INTERCEPTOR Electronics appeared in mid 1987 through US dealerships. It consists of a rivet-sealed aluminium box, 21 x 16 x 5 cm in size. In the top is a standard 6.3 mm headphone jack. The bottom of the loop has a male 6.3 mm plug which is pushed into the base unit, and can rotate 360 degrees. An on/off switch is available, plus a switchable 20 dB attenuator. A 9 volt PP3 (6LR61) type battery is clipped onto the back of the base unit

and connected. A standard SO239 connector is available for the lead to the receiver. A transparent sticker on the top of the base unit allows you to align the loop in a chosen direction, if O degrees is aligned with North.

Coverage was between 400 kHz - 2.5 MHz. The 6.3 mm jack is clever, but not a very mechanically strong way of connecting the loop to the base. The static screening on the loop reduced the noise and interference to a low level. But because the built-in amplifier is broadband in design, the Interceptor had a lot of problems with intermodulation products from mediumwave stations. The "null" is quite sharp in the mediumwave band (sometimes as much as 30 dB or 3 - 5 "S" points). But if you live anywhere near a local mediumwave AM station, this antenna is only marginally better than the ferrite rod in a portable radio. DX is buried under interference generated by the antenna itself. Away from local stations in the US Midwest for instance, it might perform better, but not in Europe or Asia.

We found the price of US$99.95, excluding shipping, rather high for the performance offered. Further details from: Interceptor Electronics Inc, Route 1, Box 439, Round Hill, VA 22141 USA. Tel +1 703 338 4905. Interceptor say they are working on new models for 1993.

ANTENNA ACCESSORIES

SP-2 ANTENNA SPLITTER

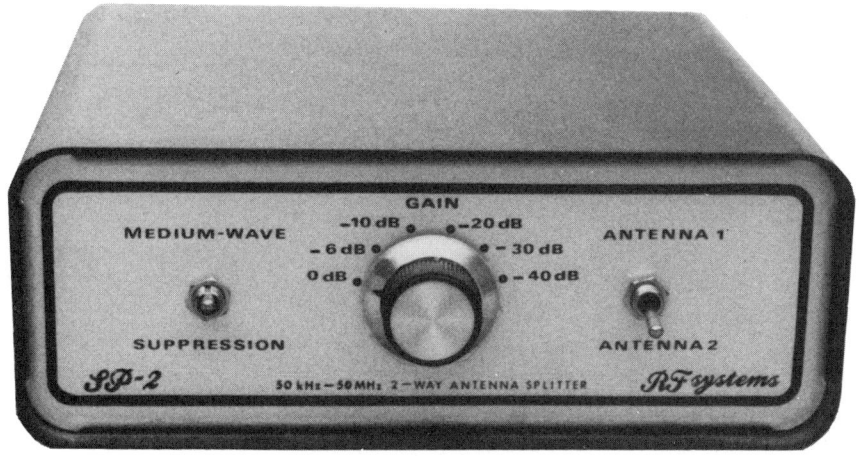

INTRODUCTION
It is quite easy to physically connect one antenna to two receivers. There are a variety of plugs and sockets on the market. However, if you try to connect two receivers in parallel to the same antenna, the results are nearly always disappointing. The reason lies in the input filters found in modern receivers. When the receiver is tuned to 6 MHz let us say, an input filter covering 4 - 8 MHz is switched in. Within that 4 MHz range, the input impedance of the receiver is 50 Ohms. But outside the range, the impedance is just a few Ohms. If you connect the inputs of two receivers together in parallel, then problems arise when you tune one receiver to 6 MHz and the other to around 13 MHz. The input of the first receiver is effectively short circuited by the second.

There is another problem. Receivers have at least one internal oscillator, and there is always some leakage which appears at the input terminals of the receiver. Since the level of these signals is sometimes in the order of that of a strong radio station, spurious signals from Receiver One are simply picked up by Receiver Two.

The received energy from the antenna has to be split in two, although since each set only notices a drop of 3 dB (in theory) the slight loss in signal strength is hardly noticeable. However, the cheaper splitters also introduce signal loss by their insertion, often as much as another 3 dB, and this is noticeable on weak signals. Cheap antenna splitters which contain simple resistors give problems since the insertion loss is highly dependent on the frequencies being tuned on both receivers, and the isolation between the sets is minimal. Thus if you try the "TV" splitters which are often marked "0 - 1000 MHz", the results are not acceptable. Even the slightly more expensive transformer types usually do not provide much isolation between the receivers until 40 MHz is reached, and higher than European UHF TV frequencies, receiver interaction is again a problem.

The SP-2 from RF Systems Inc. is a specially designed two-way antenna splitter for long-, medium- and shortwave. The SP-2 allows two receivers to be connected to one antenna, but unlike other units, this splitter ensures there is NO interaction between the receivers as a result. You can also switch between two antennas connected to this black aluminum box. The isolation between the receivers is then more than 30 dB, which ensures totally independent tuning of each radio within the frequency range of 50 kHz through to 50 MHz. This high level of isolation prevents unwanted spurious whistles due to oscillator radiation and loss of sensitivity caused by the input circuitry of the second receiver.

A constant impedance attenuator is incorporated in the SP-2. This offers a wide range of attenuation steps, i.e. 0 dB, -6 dB, -10 dB, -20 dB, -30 dB and -40 dB. Our measurements showed the steps to be well within tolerance. This allows far more flexibility when listening for weaker signals. The attenuator allows the user to reduce the level of strong incoming signals, which cause overloading or intermodulation effects in many receivers. Although it may seem strange at first, switching in some attenuation will allow weaker signals to be heard above the background noise. Too much signal is as bad as too little for most modern receivers.

There is a slight loss in signal strength by putting the SP-2 into circuit. On shortwave frequencies this was in the region of 0.5 "S" points. This is not serious at all.

MEDIUMWAVE SUPPRESSION

Mediumwave transmitters are used in many countries for domestic broadcasting. As a result, signal levels from transmitters of 50 kilowatts or more can cause blocking and intermodulation effects, even though the receiver is tuned to another part of the dial. These signals can be reduced by the SP-2's step attenuator, but this will also reduce sensitivity to signals on frequencies in the mediumwave band. The solution is to be more selective. The SP-2 has a built-in mediumwave rejection filter, which gives more than 40 dB (7 "S" points) rejection over the mediumwave range (530 - 1604 kHz) only. Since these unwanted local signals are filtered out before they reach

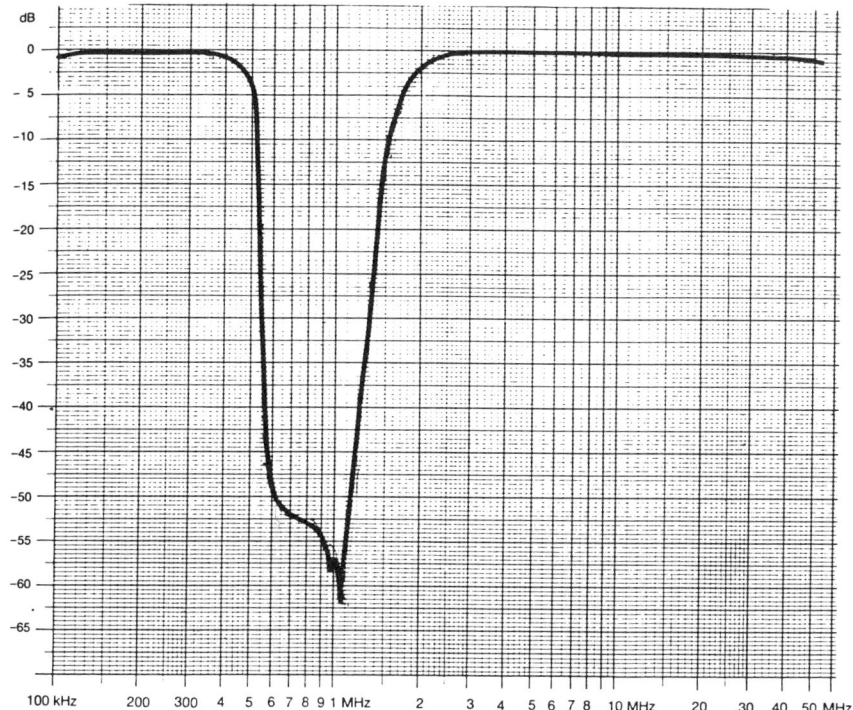

the receiver's input terminals, maximum sensitivity in the longwave (50 kHz - 530 kHz) and shortwave bands (1604 kHz - 50 MHz) becomes possible. This feature alone makes the SP-2 a worthwhile consideration even when only one receiver is connected to the unit.

TWO ANTENNAS

The polarization of received signals changes constantly., due to propagation effects. A signal received best on a horizontal antenna, can fade out and at the same moment, reception on a vertically polarized antenna improves. Listeners interested in getting the best from their equipment are therefore advised to set up two antennas, one horizontal, the other vertical. Both antennas can be connected to the SP-2. A switch on the front panel makes it possible to instant switch between antenna 1 or antenna 2. without having to unscrew any connectors. The antenna which is switched out of circuit is short circuited. This prevents any static build up which can cause damage to sensitive receiver input circuitry. The metal cabinet of the SP-2 provides an effective shield against external interference sources, such as nearby fluorescent lights.

The system impedance remains constant at 50 Ohms, irrespective of how much attenuation has been selected. This prevents loss of sensitivity loss due to reflections. The patented splitting transformer ensures the loss introduced by connecting the SP-2 is extremely low: < 0,5 dB above 3 dB

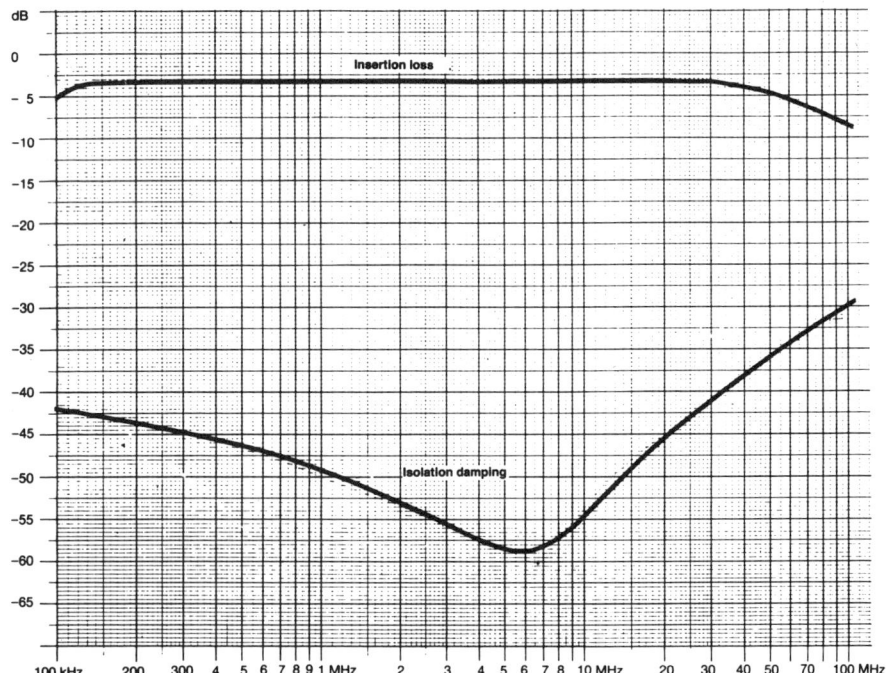

power division. The back of unit has four SO 239 connectors, and the cabinet measures 160 x 60 x 150 mm. The unit weighs 900 grams.

The SP-2 is marketed in Europe by Doeven Electronics, Schutstraat 58, 7901 EE Hoogeveen, The Netherlands. The price, excluding shipping, is 279 Dutch Guilders (approx. US$137). For the serious shortwave listener using a communications type receiver, this represents a good value choice.

PALOMAR AMPLIFILTER™

This is a simple unit designed to be connected in between the antenna coaxial and the receiver's input. The four buttons offer four functions. A 30 MHz lowpass filter removes any signals above 30 MHz. The 3 MHz highpass filter removes the signals from the mediumwave AM band which can cause cross-modulation products on shortwave reception A one step 20 dB attenuator switch is offered, plus a 20 dB wideband amplifier. You would not need the amplifier function, except in low signal areas, or when using a

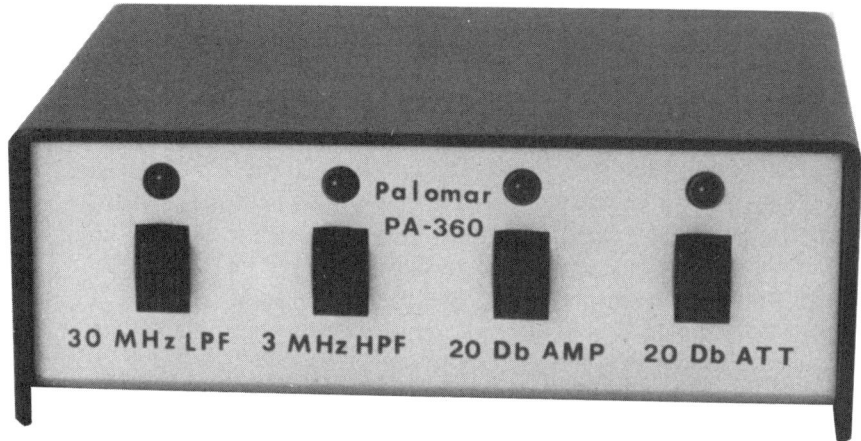

very short antenna. All this is not new technology, just traditional circuitry in one box. Bearing in mind the price of US$79.95, excluding shipping and power supply, we had expected an attenuator with more than one step of attenuation, as on the SP-2. Proper use of an attenuator can be a very useful listening aid. Further details from Palomar Engineers, Box 455, Escondido, California, 92025 USA. Tel: +1 619 747 3343.

LOWE PR-150

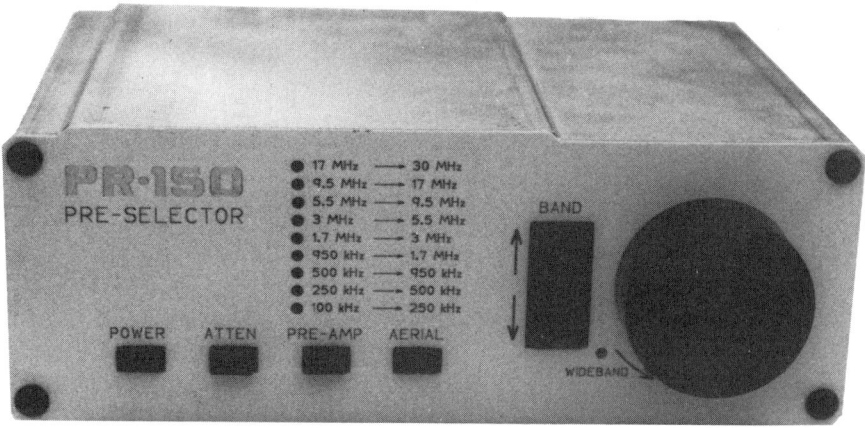

Launched in late 1992, this preselector is specifically designed to operate with the HF-150 receiver from the same company. However it can be used as a stand-alone unit with other receivers. The simpler your receiver, the more dramatic the results.

Use of the PR-150 is straightforward. You connect the PR-150 between your antenna and the receiver input. The unit requires 12 volts DC to power the logic switches and an optional 16 dB preamplifier. In the "WIDE-BAND" position the unit is by-passed. To achieve proper preselection, you first choose the range required. The preselector has 9 bands, covering portions of the radio spectrum between 100 kHz - 30 MHz. For instance, say you are interested in hearing a station on 12030 kHz. You set the preselector to the band marked 9.5-17 MHz. Then by turning the tuning knob you will find a point where the signal strength suddenly increases. This is the peak tuning position. Signals outside that portion of the dial are suppressed. In fact you may notice that the best results are obtained when the preselector is slightly off-tuned. The strength may be down but the signal sounds less distorted.

You can also attenuate signals if they are too strong, but we would have liked to have seen more steps than the single 20 dB of attenuation that is offered. Stepwise attenuation with about 4-6 steps would be ideal. For further information on the PR-150, please see the test of the HF-150 in the Receiver section.

PALOMAR CHANNEL CLEANER

This is an indoor antenna designed specifically for use with battery powered portable radios (i.e. Sony ICF-SW7600). It is based on a design by Mike Villard who describes shortwave directional loops in this edition of the Equipment Buyers Guide. The Palomar PA-420 Channel Cleaner is a commercial version of one of the loop designs.

This unit is completely passive. It doesn't try to amplify incoming signals. Instead it nulls out co-channel interference problems when two stations are beaming in from different directions. We would concur with the instruction booklet that best results are obtained when the whole unit (including the radio) is put onto a large plastic kitchen turntable. This allows you to rotate everything easily, without hand or body capacity interfering with the tuning procedure.

If you find a desired station is being interfered with by another on the same frequency, you tackle the problem by trying to reduce the signal strength of the interference. A combination of two controls on the Channel Cleaner are used to find the point where the unwanted station is "nulled" out. We found this was only really effective on fairly strong signals, usually from within a distance of 3000 km. You also need to have patience to find the null in some cases. Having found the null, the whole unit is then rotated to a position where the desired signal peaks.

The position of the "Channel Cleaner" is important. It is no good using it on any metal surface, or within several meters of a TV or computer terminal.

Palomar Engineers (Box 455, Escondido, California, 92025 USA) are a long established manufacturer of SWL accessories. The "Channel Cleaner" is priced at US$79.95 (excluding shipping) in the US. Bringing it to Europe by airmail will cost in the region of US$94.00 plus any local import duty. That price is around half what you might pay for a good portable radio, so you have to put this unit into a price perspective. The science behind the unit is sound, and the fact that it is passive means it will not overload the sensitive input circuitry of the radio. It is effective in high signal areas (such as Europe or Asia) in at least partially separating co-channel interference. The signals MUST originate from different bearings, or both signals will be nulled and the effect is virtually zero. But if you like experimenting with antennas, this compact loop may well be a worthwhile investment.

CHAPTER 5

WHO CARES ABOUT STANDARDS?

VICTOR GOONETILLEKE, COLOMBO, SRI LANKA EXAMINES A NEGLECTED PROBLEM AFFECTING THE THIRD WORLD

The last decade has seen a revolution in receiver design, to a point where the purchaser has never had so much technology for the price paid. However, the Third World SW listener is still waiting for these sets to come within his or her reach. A close scrutiny of this World Radio TV Handbook Equipment Buyers Guide shows that a good radio for international listening still costs in the region of US$300 or more.

Whilst some PTT administrations talk of the future being with SSB, the Third World (and many parts of Eastern Europe) struggle with outdated AM technology of the 1960's. Radio sets on sale in most parts of Africa or Asia do not incorporate the PLL system or digital frequency readout. Single Sideband is only to be found on amateur radio equipment. In many cases, international shortwave stations do not realize the poor quality of the receiving equipment being used in the target area.

VICTOR GOONETILLEKE

Most domestic radio sets have FM, MW and usually cover the SW range between 2 - 22 MHz in two or three SW bands. The best radios cost in the region of US$100, roughly the average monthly income in city areas. To avoid import restrictions, many receivers are locally manufactured, even if they carry a brand name associated with a European or Japanese company.

These locally produced receivers have adequate sensitivity, but the dial calibration is so rough that fine tuning on shortwave is difficult. Scales are often out by as much as 150 kHz. Since the circuitry is based on a single conversion design, problems with "ghost signals" or images is a major problem. Cross modulation is rampant the moment anything other than the telescopic antenna is used. The simple design makes it quite difficult to separate two strong signals even 10 kHz apart. Such receivers are adequate to tune in the better known international broadcasters who operate relay stations. But if the signal strength is anything less than an S-4 on the SINPO scale listening is very difficult.

What then is the Third World listener waiting for? In very simple terms he or she is waiting for a continuous coverage PLL receiver with a digital readout, capable of tuning down to 1 kHz segments. The minimum coverage should be 4 - 22 MHz, plus FM and MW for local stations. In 1990 we saw the introduction of one PLL receiver for US$50.00 in the US. If another US$20.00 could be spent on improving the image rejection of the receiver, then the result would begin to look like the ideal receiver for the Third World. Whilst SSB capability would be a useful extra for the hobbyist it is not (yet) essential, and features such as scanning and memories are luxuries which should not be included if this means compromising on receiver performance. The manufacturer would have to be prepared to allow local production or assembly, otherwise import duties between 12 & 100% will price the radio out of the market.

It seems surprising that more use is not made of solar cells as a power source, perhaps charging a nickel cadmium battery to allow for night-time reception. In Africa, especially, this is not a gimmick, when batteries are extremely expensive and in short supply. Power cuts are frequent, especially in times of crisis when you need the radio most for information. There were studies by UNESCO a few years ago to produce a solar powered set, but it doesn't appear to have resulted in a marketable product.

NICE PANEL..SHAME ABOUT THE SOUND!

Having been a keen international radio listener for more than 20 years, as well as an amateur radio operator and DX club editor , I am amazed that stations themselves have not done much to promote better receiver standards. If the European Broadcasting Union or the Asian-Pacific Broadcasting Union can meet to discuss technical standards for high-definition TV, is it too much to ask for a international minimum standard on shortwave receivers?

Scanning the reviews in this book, it seems that receiver manufacturers are not being given any guidance by stations. Why else would manufacturers insist on using bandwidth filters 12 kHz wide, when most shortwave stations don't broadcast audio frequencies much above 6 kHz? Broadcasters forget that the link between them and the listener is not completed by just blasting the program out of high-gain transmitting antennas! shortwave radios have more gadgets on them in the last five years, but what percentage of the cost goes into making the vital receiver portion. I suspect it is less and less.

Whilst global TV operations such as CNN may give the managers of shortwave stations the feeling that the end is near, it will be decades before TV can sort out its distribution problems in developing countries. Even local re-broadcasting can lead to questions of who has the final say.......sensitive material can easily be cut by those in charge of the domestic transmission facilities. Shortwave radio remains a relatively cheap, portable medium. It remains an essential news source in places where domestic radio is censored, and TV is just transmitting often, irrelevant, entertainment programs. For the majority of the Third World, and of course the newly emerged free nations of Eastern Europe, international radio is still an important part of daily life. It is more than just an enthusiast's exotic hobby.

For the many millions of radio listeners in the Third World the so-called semi-professional receivers like the Kenwood R5000, Icom R72, NRD-535 are dreams which rarely ever materialize. For thcm, and that includes the author, the ICF-2001D has been a dream almost come true, but it is still wishful thinking for most of us. Will receiver manufacturers and broadcasters ever get together to put the essential features into one, affordable, plastic box? Let's hope so.

CHAPTER 6

COMBATING INTERFERENCE IN SHORTWAVE RECEPTION WITH COMPACT INDOOR DIRECTIVE ANTENNAS

BY O.G.VILLARD, JR. SRI INTERNATIONAL, MENLO PARK, CALIFORNIA 94025, USA.

Enjoyment of shortwave international radio broadcasts is not infrequently degraded or totally spoiled by the presence of interference, either local or distant in origin. The latter is becoming more common as the number of broadcasts and their power levels increases. One way to reduce interference is to use antenna directivity, either at the transmitting or the receiving end. Although directivity has been exploited by broadcasters to the extent permitted by economic considerations, it has seen comparatively little use by shortwave listeners for three reasons: Firstly, outside directive antennas are too bulky and costly in most situations. Secondly, the principal directive antenna compact enough for indoor use (the classical tuned loop) has unsatisfactory performance at shortwave frequencies and finally, past experience with HF direction finding in the presence of nearby reflectors suggests that any attempt to separate stations indoors would be bound to fail.

New developments, however, have changed this picture completely. A variety of directive antenna designs, compact and sensitive enough for indoor use, are now currently available. The unsatisfactory performance of traditional loops at short waves has been overcome, and extensive experience with indoor reception has shown that discrimination between two signals on the same channel is possible, even when accurate direction finding is out of the question. The result is a new category of directional antennas for shortwave, whose small size (made practical by the fact that broadcasters are using high power transmitters) enable them to be located indoors close to the receiver where they are easily accessible for adjustments. They can be thought of as being the longer-wave counterparts of indoor antennas (some active, some passive) sold for FM and TV application as improvements on the traditional "rabbit ears." The HF antennas operate by generating an adjustable null direction in their response pattern, rather like the null of a notch filter in a receiver pass-band.

The following is intended for listeners who are fed up with interference and wish to take positive steps to reduce it. It is also intended for serious DXers (individuals with a more technical interest in shortwave listening) who will find that nulling out a strong station on a given channel often uncovers weaker ones whose presence was previously unsuspected. All listeners will also find that even rough directional information helps to identify unknown stations.

BASICS

Local interference reaches the listener by travelling along the surface of the earth. This is called "ground-wave" propagation. Signals of distant origin—the fading kind—reach the listener by reflection from ionized layers in the upper atmosphere. This is known as "sky-wave" propagation. Both types are illustrated in Figure 1. Of the two, sky-wave interference is the more difficult to get rid of, so most of the following discussion will be devoted to it.

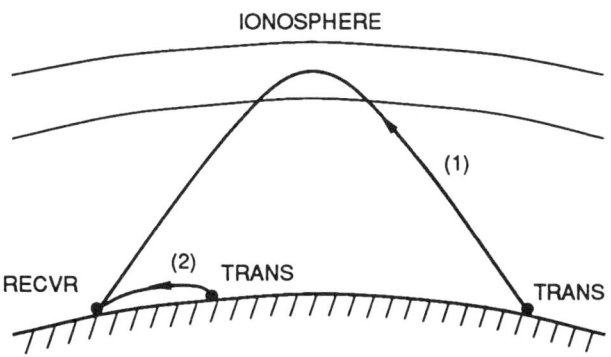

FIGURE 1 Possible transmission paths for interfering signals.

The first schemes to be discussed are simple, both to build and operate. More elaborate and better-performing arrangements are described later. Thus the listener can choose the techniques that best fit his or her requirements. The simplest arrangements take advantage of the self-contained nature of small battery-powered receivers, and require no external electronic components at all. They give a front-to-back response ratio with sky-waves of roughly 15 dB and no adjustments are needed either when changing frequency band or station. A somewhat more complicated approach is required in the case of power-line-operated receivers with coaxial inputs. However, in that case, the average desired-to-undesired sky-wave signal ratio can exceed 20 dB. The same ratio in the case of ground waves can be very much higher.

In practice, a discrimination of 20 dB appears adequate to separate two broadcasts of comparable strength on the same frequency in spite of ionospheric variability and fading. All that is required for signal separation is to establish a null in the direction of one broadcast or the other. The listener then has a choice of two individual programs instead of one unintelligible mixture.

Attenuating one signal by 15 dB may not seem like a major victory considering that broadcasts may be 60 dB or so above the noise. However, interfering signals are often of strength comparable with the desired ones; 15 dB can make the difference between understanding the gist of a program and failing to understand much of anything. In principle, several of the antennas are simple enough that they could be incorporated in receivers at the time of manufacture. If such devices were used in sufficient number, it is possible that more stations could be squeezed into the existing channels than at present. Since the antennas can also attenuate sky-wave interference in the medium frequency broadcast range, their use could influence antenna design and night-time power assignments in that part of the spectrum as well.

NULLING VERSUS SEARCHLIGHTING

Full-size directional antennas or beams used at professional receiving stations behave rather like searchlights. They increase the strength of the desired station in relation to the strength of signals coming from other directions. Searchlight antennas small enough to be suitable for use indoors, however, are too complex. Great simplification results when a receiving antenna is designed to attenuate one undesired signal at a time. This is not a serious handicap because the probability of encountering two stations of comparable strength interfering on the same frequency is considerably greater than that of encountering three or more. Of course, nulling antennas can take out two stations at once which lie in roughly the same direction, thereby uncovering a desired third transmission on the same frequency. Because of an indoor antenna's small size and ready accessibility, null directions can be significantly shifted manually by the operator in fractions of a second...something that is not possible with a full-size beam. This is useful when stations come and go, as in the case of amateur radio "roundtable" discussions.

DEFINING PERFORMANCE

In the following discussions it is assumed that (a) there are no more than two stations on the same channel, (b) stations are separated in azimuth by 45 degrees or more, (c) the antenna is located inside a building where reception is reasonably good and (d) ionospheric conditions are normal.

"Performance" in separating stations can be defined as the extent to which a given signal on a given frequency is attenuated when the "antenna" is in the minimum-strength or null position, in comparison with the direction that gives maximum response. If two interfering stations are reasonably different in direction, the probability is low that an adjustment which nulls out one signal will appreciably affect the other. For this reason, one station by itself may be used for test and tune-up purposes.

The best nulling antennas have adjustments which permit them (momentarily at least) to reject a given sky-wave signal by truly spectacular amounts, around 40 dB when conditions are good. The adjustment is a balancing act between an "amplitude" control and a "phase" control. However, due to the changing ionosphere, the null will not generally remain that deep for long. Often after half a minute or so, some readjustment becomes necessary. The action is not unlike balancing a bridge whose parameters are slowly varying by small amounts with time. If no readjustment is made, and performance is averaged over 10 to 20 minutes, a rejection ratio of 20 to 25 dB can be expected. (In the case of ground waves, which are much more stable, no retuning is necessary, and a 40 dB rejection figure will hold for tens of minutes.) Nulling antennas have been dubbed "spatial notch filters" by analogy to filters whose action depends on the frequency of the undesired signal. The spatial devices, however, work well even when the frequency of the interfering signal is exactly the same as that of the desired signal. Of course, when frequency filters can be used, their rejection adds to that of the spatial kind, and the sum of the two can be an impressive number of decibels.

Although indoor nulling antennas can produce gratifying improvement irrespective of whether the undesired signal arrives via sky- or ground-wave, it is also true that they will not result in major improvement in all sky-wave cases. Reasons include the effect of scattering from nearby conducting objects such as buildings (particularly those blocking the line of sight); scattering from the earth when a distant transmitter's beam is directed away from the great circle path leading to the listener; the presence of signal components at high vertical angles (between 45 and 90 degrees) and (occasionally) ionospheric disturbance. All these have the effect of converting a clean deep null into a wide and blurry one. As a rule of thumb, it can be said that at any given time and broadcast frequency band on the average perhaps 50 % of the sky-wave signals can be significantly reduced in amplitude by nulling procedures. The exact figure depends on the chosen international broadcast band, time of day, listening location, and the desired and undesired transmitter locations. Usually the best nulling is observed with reasonably distant slowly fading transmissions reaching the listener via one ionospheric reflection. Performance with amateur radio signals is surprisingly good, presumably because scattering is less noticeable with transmissions of lower power.

In- or outdoors, a radio and/or its antenna can be mounted on an inexpensive plastic kitchen turntable (obtainable at most hardware stores) for easy rapid rotation. (Sheets of slippery paper or plastic make a satisfactory, though not very elegant, substitute.) Since continuous rotation is not required, the antenna cable to the receiver (when used) can be allowed to wind up, thereby avoiding the need for a slip connection.

RADIO WAVES' INTERACTION WITH BUILDINGS

It is assumed in the following that the listener will do the majority of listening indoors, although antenna directionality will in general be better outdoors. Incident radio waves are well and truly disturbed by the typical house.

Wood frame houses are the most transparent; reinforced-steel ones the least. Brick and masonry structures lie in between. Fortunately, most buildings have windows, and glass admits short radio waves almost as well as light. HF compact directional antennas when placed near a window in an otherwise impervious (electromagnetically speaking) building, can often separate stations even if the direction indication is hopelessly inaccurate. One reason is that radio signals illuminating one side of a building excite "creeping" waves which are guided around the building to the opposite side. These waves retain some semblance of their original directionality, thus allowing discrimination.)

It is always desirable to be able to move a radio being used indoors so as to find the best room and the best location within a room. Often a change in position of as little as a meter makes a noticeable difference. Energy which leaks in via windows or walls is supplemented by some which is guided in by conductors such as telephone lines, electrical wiring, and plumbing vents. Near these one-wire transmission lines, and especially near their ends, (for example, at telephone instruments, TVs, table lamps, etc.) the guided energy reappears in the form of local electric fields which, luckily, fall to the background level within no more than a meter or so. Exploration with a portable radio will quickly show where such fields are to be found, and the hot spots should, of course, be avoided wherever possible. These local guided-wave electric fields represent an appreciable share of the total RF energy indoors. It turns out that anti-interference antennas designed for low impedance, such as loops, are much less sensitive to them. They give noticeably more accurate direction indications than their high impedance counterparts, the whips.

In addition to serving as a guide for waves travelling along its exterior, a typical building reflects some incident energy, absorbs some, and admits the rest. The latter component weakens rapidly as it moves inside. If a complete choice of indoor locations is available, it is clear that any kind of a receiver and indoor antenna ought to be on the building side closest to the desired station and farthest from the source of interference. Use of the building as a shield in this way is partially spoiled by the guided-wave effect, but fortunately not completely.

DIRECTIVE RECEPTION WITH WHIP ANTENNAS

The polar pattern of the whip antennas provided with most portable receivers is normally considered to be non-directional and consequently of little use in rejecting interference. (Part of the problem is the lack of symmetry between the vertical whip and the horizontal receiver body.) But when receiver and whip are oriented in the proper near-horizontal fashion, their response can be made to resemble that of a horizontal wire dipole in free space. For this to occur they must be separate from other conductors and there can be no direct wire connections to the receiver. Directionality of an ideal horizontal wire dipole resembles that of a loop, which is the traditional means for obtaining indoor directivity at HF. Full-size indoor horizontal wire dipoles are not popular with shortwave listeners because of their awkward length. However, small dipoles such as the ones to be described can be easily manipulated.

A small dipole formed of a receiver and its whip has powerful advantages: (1) the response is independent of frequency (as is that of the whip by itself) requiring no tuning or adjustment when changing international broadcast bands or stations, (2) no additional electronic parts are required, (3) directivity can be adjusted for either unidirectional or a bi-directional response, and (4) either sky-wave or ground-wave interference can be attenuated.

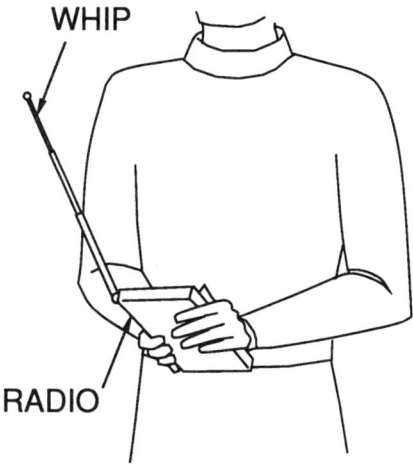

The simplest directive arrangement is shown in Figure 2. The receiver and whip are held close to the chest, with the whip extending radially outward. Body and whip are then turned to the position at which an undesired signal is weakest. (This requires that the incoming vertical-plane mode structure not be too complex, and the signal(s) not too strong.) The whip then points roughly in the direction of the station to be eliminated. In finding the null position, the whip should be elevated or lowered in the vertical plane by moving the receiver so as to establish the position of weakest signal. Adjusting the length of the whip may help. The received signal should markedly increase when the listener rotates him- or herself so that the whip moves from the null position to a different azimuth but at the same elevation.

Figure 2: Simplest directive system. For best undesired-signal rejection, the listener adjusts whip in azimuth, by rotating body, and in length and elevation by raising and lowering whip. (Indoor results are location-dependent: be sure to try several.)

From the standpoint of a small portable radio being operated indoors, the screening effects of metal in the building means that most houses turn out to be filled with radio "shadows" not unlike the holes of a Swiss cheese. However, the indoor shadow of a desired station does not perfectly coincide with that of the undesired signal (or vice-versa) since the signals usually emanate from different locations. It is this fact which permits stations to be separated by locating a small radio "probe" in the right place, just where the shadow of the unwanted station is deepest. At that point the desired station will be received best. But the success of this method depends on the metal clutter in the vicinity of the receiver. It may work well, but with bad luck the shadows may almost coincide and separation will be poor. It is purely a case of trial and error.

Estimating changes in signal strength is not easy with modern radios having excellent automatic gain control action. But, when two stations of comparable strength are interfering on a common frequency, the effect of a properly adjusted antenna should be noticeable. When only one fairly strong station is audible, strength changes associated with antenna directivity are best estimated by noting the change in background noise level. Retracting the whip to the point where the background noise appears will aid this process.

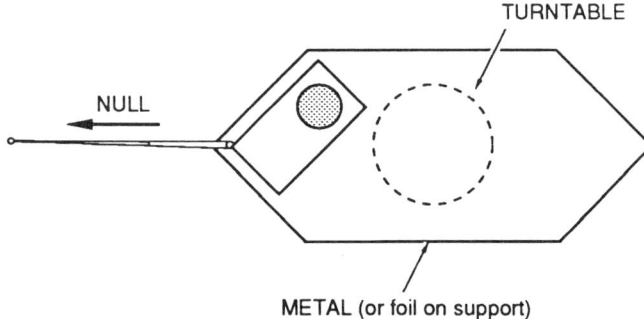

Figure 3: Orthogonal-Probe Directive Antenna (OPDA). Only requirement is a piece of metal or metal foil, plus a turntable. Uni- or bi-directional pattern. No antenna tuning required.

In general, operation can be improved by regularizing the fields surrounding the receiver. A number of advantages are conferred by placing it on a sheet of metal as shown in Figure 3. The whip should extend outward from the center line. The dimensions of the metal sheet are not critical, provided that the generally diamond shape is retained. The arrangement has been named the Orthogonal-Probe Directional Antenna (OPDA for short).

HOW THE OPDA WORKS

Radio, plate, and whip are oriented so that the whip is orthogonal with respect to the field lines of the incident wave. Little or no voltage is then induced on the whip with respect to the plate and the radio (which is electrically part of the plate). In that one whip position, the received signal strength is very low. At most others, it is considerably stronger.

Since most shortwave broadcasts arrive at an angle of a few degrees above the horizontal, tilting the whip upward by that amount helps satisfy the orthogonality requirement. Then when the assembly is rotated 180 degrees, the whip is no longer at right angles to the electric field lines and a comparatively strong signal is received. In practice the whip is adjusted as to tilt and length to minimize the strength of an undesired signal. Both the sheet and receiver are rotated in azimuth as one unit to help in this process. When a null is found, the whip does not necessarily point in the azimuth of the distant station, since it must point in the direction perpendicular to that station's electric field lines, even when they are distorted by local effects. Null positions found indoors in this way are often unstable with time, so that occasional readjustments may be required. Since some locations are better than others in this respect, experimentation is recommended. Performance outdoors will be much more stable. A further advantage of the metal plate used with the OPDA is that when signal levels are high and performance is degraded by direct signal leakage into the receiver, it is possible to retract the receiver's whip and to move the receiver further toward the center of the metal plate (space permitting). This improves shielding and reduces overloading.

APPLYING THE WHIP-LOOP PRINCIPLE (CSWL)

An alternative approach for achieving directivity is to use the receiver whip as a non-directional vertical antenna to which the output of a standard bi-directional loop is added. By adjusting the proximity and nature of the coupling, the resulting pattern can be made unidirectional. (Such a pattern is considered preferable to bi-directionality because the likelihood of accidentally attenuating a desired signal is smaller.) A loop for use with a compact receiver should have a size in proportion to that of the receiver's whip. If the loop is to have a small area, it must be resonant. These considerations lead to the Close Spaced Whip Loop, or CSWL, of Figure 4. A circuit diagram is shown in Figure 5. Note that the whip is located in the plane of the loop. Addition of a resonating capacitor to the loop gives useful amplification and permits easy adjustment of the phase of the voltage developed across the loop. Voltage on the side of the loop nearest the whip is coupled to the whip by "stray" capacitance—no conductive connection is required. Such capacitive coupling automatically gives the coupled energy essentially the correct phase for a unidirectional response. The small correction needed is easily provided by the tuning capacitor. Amplitude of the coupled voltage can be controlled by a shunt variable resistance which alters the loop Q. Note the position of the loop with respect to the whip (upstream), and the fact that the upstream side of the loop is coupled via the capacity of a plate to the receiver's chassis ground. These details happen to be important.

Operating the device is straightforward. First, the resistance is set to its highest value. Next the desired station is tuned in. Then the tuned circuit is

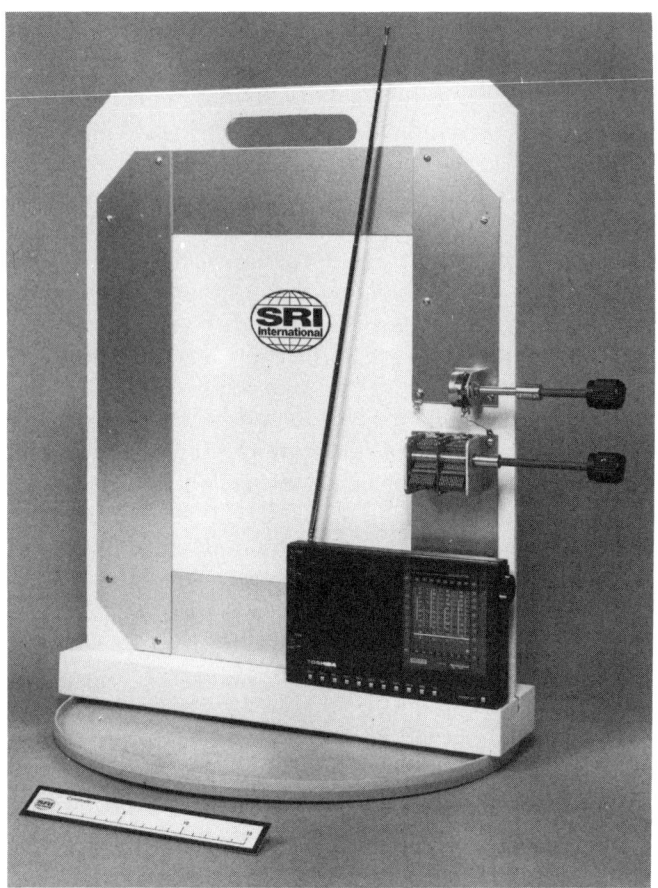

Figure 4: CSWL—A simple means for adding directionality to a battery-powered portable with a telescopic whip. Useful with both sky and ground waves. The capacitor shaft points toward the signal to be eliminated.

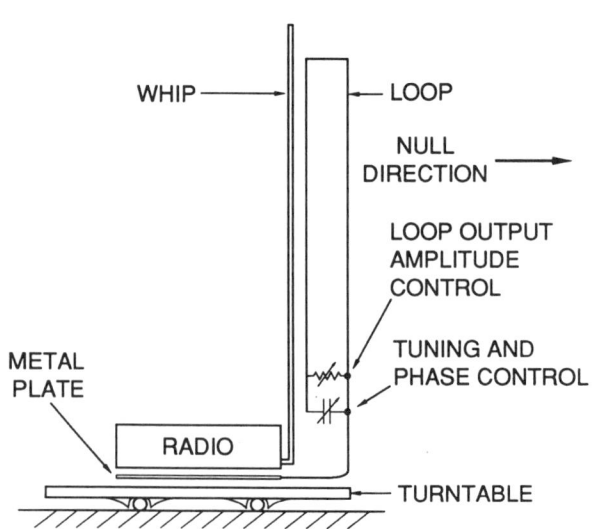

Figure 5: Schematic of CSWL. Note: there is no electrical connection between the plate and radio. (Plate and radio are insulated.)

adjusted to the same frequency, whereupon a marked change in signal amplitude should be observed. This might include both a signal increase and a dip as the circuit is tuned from one side of resonance to the other. If the dip is slight, resistance should be decreased until it is as deep as possible. At that value of resistance, it is normal for the resonant peak not to be noticeable. (Of course, if the resistance is too small, coil Q will vanish and tuning will have little or no effect.) Turntable position must be adjusted along with the tuning control for greatest dip depth. Rotating the turntable may not be necessary if the direction of the station to be nulled is known in advance. (All this may sound complicated, but it is quickly mastered.) Headphone use is not recommended because the connecting wire tends to spoil the null by acting electrically as an extension of the radio. The same problem arises when the radio is operated from the power line. (This is discussed in the next section.) If a headset must be used, the listener should try to locate himself downstream from the undesired signal, and should try to keep the wire at all times as nearly perpendicular as possible to the electric field lines of the incoming wave to be nulled. (This is the principle which is used to minimize whip pick-up in the null position with the OPDA.) It will be noted that with the CSWL the null is formed when the station direction is in the plane of the loop, instead of perpendicular to it. This eliminates the poor performance with sky-wave signals which has plagued conventional loop direction finders for years. A commercial ver-

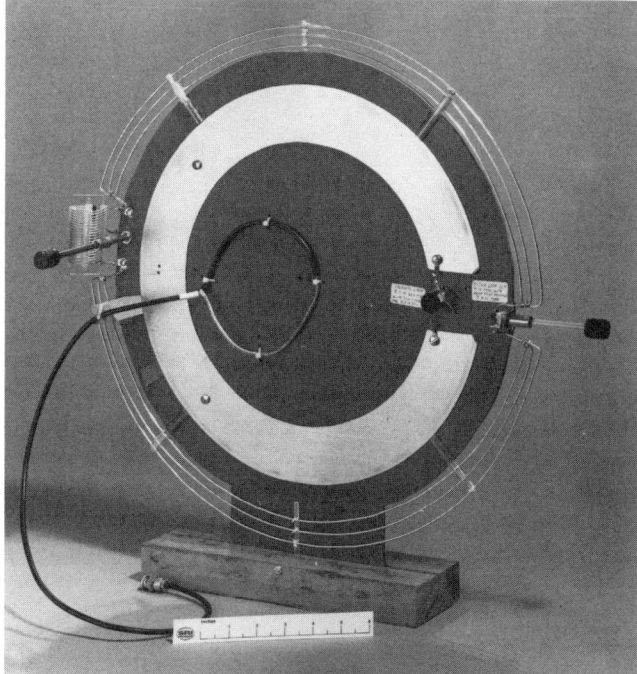

Figure 6: Coplanar Twin or Tri-Loop (CTL): needed for receivers with coaxial feed. Central widestrip loop is the primary resonator. Outermost loop responds to electric field like a whip; it must be retuned when changing channels. Overall CTL performance is the best yet attained.

sion of Figure 4 is available from Palomar Engineers, Box 455, Escondido, California, 92025, USA and can be ordered as the CHANNEL CLEANER, Model PA-420.

DIRECTIVITY FOR RADIOS WITH COAXIAL INPUTS

Many communication receivers have a coaxial RF input and operate from the mains power. Coaxial feed requires an antenna that is a separate entity, supported by itself on the turntable. The essential problem in designing compact directional devices of any kind is preventing RF energy picked up on the feed line and on the radio itself, from travelling back to the antenna via either the outside of the coaxial feed cable or by any power supply wires (in the event that a preamplifier or powered preselector is being employed). The loop, or whatever directive device is used, can only make an accurate determination of direction when the ambient field surrounding it is undisturbed. Energy which has reached the loop by travelling "antenna-style" along conductors such as the braid of a coax, gives rise to a disturbed resultant field which prevents the loop from working out the true ambient-field direction. Conductors of appreciable length behave like antennas and one-wire transmission lines rolled into one. Whenever a loop is effectively a part of such a conducting path, the conveyed-in energy can be expected to upset operation. In principle, currents on the outside of coaxial lines can be suppressed by the use of ferrites. Unfortunately, these are low impedance devices. They interact well with currents but not so well with voltages. A preferable technique for achieving isolation is to avoid direct connection to a low-impedance loop and to extract energy from it by

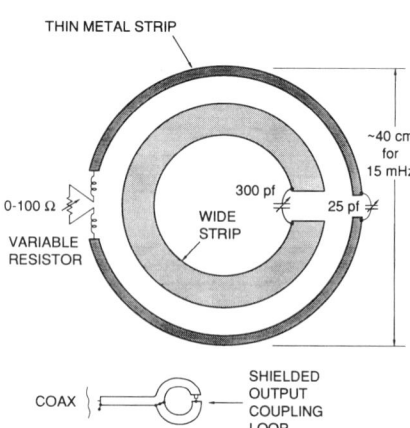
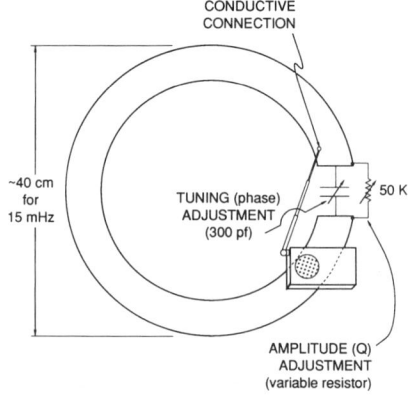

Figure 7: Schematic diagram, CTL antenna. Output coupling loop is located inside the main loop as in Figure 6.

Figure 8: Schematic, showing central loop of Figure 6 with small receiver connected directly to it for sensitive operation. Capacitance from loop to receiver case completes the circuit.

means of a shielded and balanced coupling coil as in Figures 6 and 7. The amount of energy transferred by capacity is then very low because both loop and coil have a low impedance.

THE COPLANAR TWIN LOOP (CTL)

The considerations just discussed have been incorporated in a design originally called the Coplanar Twin Loop (CTL). However, when the output is derived from a third loop, as is done in the following, the CTL really becomes a "Coplanar Tri-Loop." Directivity is determined by the single-turn main loop which, for low impedance, can conveniently be made of a wide metal strip as shown in the figures. A standard broadcast-band capacitor will tune it over a wide frequency range. These can be bought at amateur radio stores, or salvaged from an old radio. Note that no shielding is needed. Because of the planar construction, the antenna responds only to vertically polarized signals. To make the response of the loop unidirectional, the output of a whip (or an electric field sensor) must be added in the proper phase, and with the proper amplitude. Again, conductive connections of any kind must be avoided to prevent the antenna effect. To add the necessary whip voltage, a larger loop of proper design can be inductively coupled to the main loop. This outer loop is given a high impedance by using a small tuning capacitor and by incorporating a loading coil at the center, thus giving a high L/C ratio. (The coil should resonate at the desired frequency with perhaps 15 pF of capacitance.) See Figure 7.

As a result of the high impedance, the combination of loop and loading coil behaves both as a whip and as a loop—the former action extracts the desired reference voltage and the second provides inductive coupling to the center loop. Inductive coupling automatically provides the needed 90-degree phase shift, which is just as effective as the capacitive coupling in the Close Space Whip Loop described earlier. The amount of outer-loop energy added to that of the main loop can be varied by adjusting the outer-loop Q by means of a variable resistor. More details will be found in References 1, 2, and 3 at the end of this article.

Operation is in fact quite simple. The main loop is tuned to resonance at the desired broadcasting band and then left alone. The innermost loop (which connects to the receiver) requires no tuning at all. The outermost loop must be tuned to the signal—the correct setting is that which gives best rejection of the unwanted signal. The assembly is rotated until maximum rejection of an unwanted signal is achieved. Resistance control is finely adjusted to obtain the greatest null depth. The plane of the loops should then coincide with the direction to the station, and at that azimuthal setting the null obtained will then be the most stable. No further adjustment is required until the station frequency is changed. With sky-wave signals having low vertical angles of arrival, performance is impressive. At higher angles, performance worsens, but it is as good as can be achieved with any antenna working on the whip-loop principle.

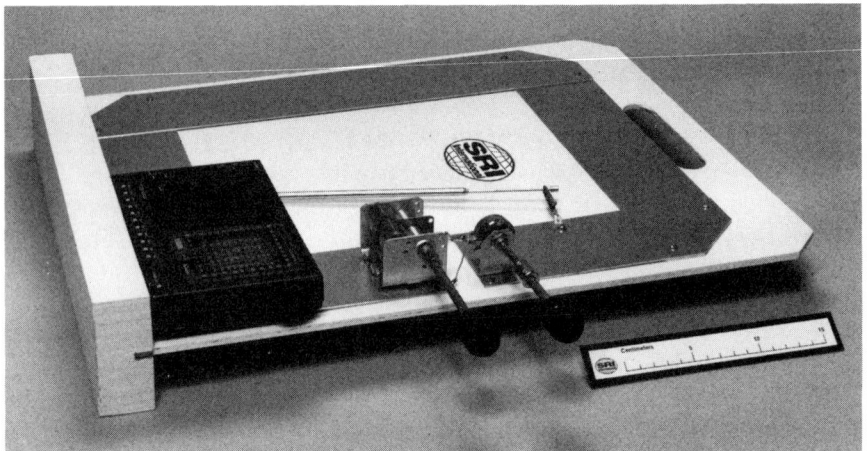

Figure 9: When operated nearly horizontally with its receiver, the main loop of Figure 8 (called the LILA), attenuates ground waves irrespective of direction while retaining sensitivity to skywaves. Nulls in excess of 40 dB are attainable by careful tilting. (Resistance is not needed for this application.)

THE LOW-IMPEDANCE LOOP ANTENNA (LILA) BY ITSELF

It turns out the low-impedance main loop of a Coplanar Tri-Loop (Figure 7), proves to be an attractive receiving antenna by itself. It has been named the LILA. For especially sensitive reception a whip-equipped radio may be directly connected to the LILA as shown in Figure 8, in place of the output loop needed for coax- equipped receivers. The radio's whip is electrically connected to one side of the loop; capacity between the radio's case and the other side completes the circuit. When connected in this way, the loop is considerably more sensitive than the radio's whip and displays the well-known noise rejecting ability of vertical loops. As a result of its high sensitivity and low impedance level, the LILA can also be operated in a near horizontal position near to the ground (about 1 m), where it will be found to reject ground-wave interference with little loss in sensitivity to downcoming sky-waves. See Figure 9. The best signal-to-interference ratio in the case of one ground-wave signal (or several such signals superimposed) can be obtained by tilting the loop slightly out of the horizontal. The loop has useful discrimination against all ground-wave signals essentially independent of direction of arrival, which makes it particularly useful when interference can be expected from a number of directions at once. An example would be a series of leaky insulators on a power line running past the receiving location. It is also ideal for situations where interference may appear at random in unpredictable directions. (For example, from sparking pantographs on electric trains or trolley-buses in a city.) True, the loop must be tuned, but the tuning is uncritical. Because of the low impedance, operation is little affected by closeness of the human body. Additional information on construction and use of the LILA is given in Reference 4, including the home constructed equivalent of a commercially manufactured variable capacitor.

SKY-WAVE REJECTION WITH THE LILA

Unidirectional reception of sky-waves as well as ground waves, with performance comparable to that of the CSWL, is possible with the above low-impedance loop if a battery-operated receiver is located close to it but not actually connected to it as in Figure 10. Both receiver and loop are in the vertical position. Because of its low impedance the LILA has relatively small electric fields associated with it; however, their magnitude is nevertheless appreciable because of loop size and capture area. According to one view, voltage appearing across the main loop's capacitor couples capacitively through space to the radio's whip, in essentially the same way as in the CSWL. It is also possible to think of the LILA as generating a direction-dependent null region in space—a "shadow"—at the position of the receiver's whip. The receiver then acts simply as a probe which registers a very low output when encompassed by the "shadow." Tuning is simpler than with the CTL, because there is only one loop to resonate. Because of its low impedance, a LILA can be tuned over a wide frequency ratio so that bandswitching is normally not necessary. The tuning procedure once the station has been selected is essentially the same as that of the CSWL, above.

It can be said that as a device, the LILA offers (1) very sensitive reception, particularly when vertical; (2) direction-independent ground-wave rejection when horizontal (Figure 9); (3) unidirectional reception of sky-waves (Figure 10)—for this, resistance control of Q is important; and finally, (4) it can readily be converted into a CTL simply by adding an outer loop. Thus the LILA makes a good general-purpose receiver accessory, usable for improving performance when not needed for interference reduction.

A professor emeritus of electrical engineering at Stanford University, and holder of the amateur call W6QYT, the author will be happy to

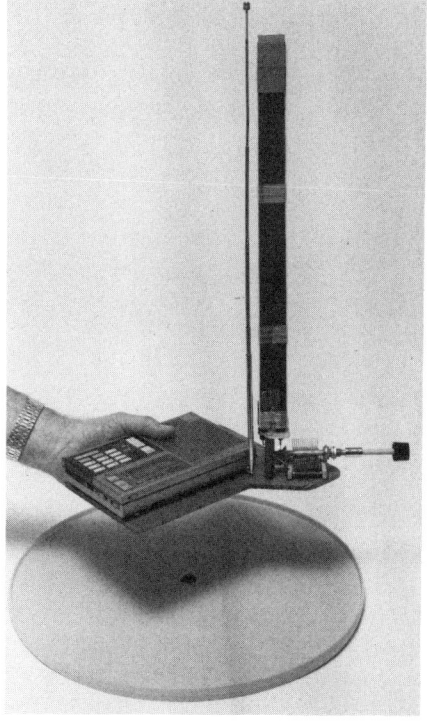

Figure 10: If the LILA is operated vertically close to, but separated from a small reciever, a unidirectional sky-wave response similar to that of the CSWL can be obtained. For this, the variable resistance of Figure 8 is essential.

respond to readers' questions on shortwave loops if they are accompanied by a self-addressed envelope and an International Reply Coupon.

REFERENCES

1. O. G. Villard, Jr., "Interference-Reducing Antennas for shortwave Broadcast Listeners," IEEE Transactions on Broadcasting, Vol. 34, pp. 159-166 (June 1988).

2. O. G. Villard, Jr., "The Coplanar-Twin-Loop Antenna," QST (American Radio Relay League magazine), Vol. 72, No. 9, pp. 29-35 (September 1988).

3. O. G. Villard, Jr., "Portable Unidirectional HF Aerial for Reducing Co-Channel Multihop Sky-Wave Interference," Proceedings of the Fourth International Conference on HF Radio Systems and Techniques, April 11-14 1988, pp. 141-144, Institution of Electrical Engineers, London, England (1988).

4. G. H. Hagn, O. G. Villard, Jr., C. A. Hagn, and M. J. Toia, "The Wide-Strip Horizontal Loop Antenna (HLA): An Effective Solution for Ground-Wave Interference to shortwave Reception," Proceedings of the Fourth International Conference on HF Radio Systems and Techniques, April 11-14, 1988, pp. 145-150, Institution of Electrical Engineers, London, England (1988).

CHAPTER 7
IS THE COLD WAR ON SHORTWAVE FINALLY OVER?

A VIEW FROM CENTRAL EUROPE
BY OLDRICH CIP

Over the years I've had the opportunity to interview the pioneers of radio in Czechoslovakia. Despite political restrictions at the time, I've also been fortunate to maintain contacts with international shortwave personalities, like Arne Skoog of Radio Sweden, Jens M. Frost, the former editor of WRTH, and even with O. Lund Johansen, the founder of World Radio Handbook during the last years of his editorship of

OLDRICH CIP at Radio Prague

that publication. Time flies, however, and I realize somewhat sadly that I am becoming an old-timer myself. Yet as the time passes I feel even more strongly that there is a need to secure a period of rational post-Cold War development for shortwave broadcasting.

WAR OF THE WORDS!

I became interested in radio as a schoolboy. I remember very vividly what it was like to listen on short waves in the period immediately after the Second World War. My brother and I had set up a simple two-valve receiver from the German Wehrmacht war-surplus parts. We were astonished with the clear reception of American and Canadian stations as well as those from other continents. By the start of the 1950's the shortwave medium began to change dramatically. During the decades of the Cold War enormous investment was made in shortwave broadcasting facilities on both sides of the so called "Iron Curtain".

The Soviet Union, and the countries under its sphere of influence, set up complete networks of jamming stations. At the same time, broadcasting stations in the West started building more and more high-power transmitters with the aim of getting their signals through the deliberate interference. A shortwave "arms race" became a reality.

Many Western shortwave broadcasters started to use large numbers of transmitters all carrying the same program in the hope that at least some of the signals would be heard. The adoption of this "redundancy" strategy was not the only devastating outcome of jamming. It also affected signals on frequencies which were well outside the intended target. This effect, which became known as "Third Party" jamming, made a considerable part of the spectrum totally useless, especially in Europe. Until recently, little was published about reception conditions in Central and Eastern Europe during the period when jamming was extensive. In fact reception conditions were the worst in the bigger cities which had their local jamming sites. Here in Prague, for example, signals of jammers with the well-known "Z3" code identification were capable of covering an entire broadcast band with several intentional - plus several spurious - signals on some occasions. The jamming seems most effective on rainy days, probably as a result of improvement in the antenna radiation due to better soil conductivity.

Despite this, however, jamming was never 100% successful, even in the cities. Reception of foreign programs was usually much easier out in the countryside. Time periods around local sunset were usually best time for reception: Long distance high-power jammers, mostly located in the Soviet Union,

Radio Prague sticker

were already skipping over Central Europe, because they were operating in darkness conditions. However, the same shortwave bands were still usable for stations with transmitter sites in Western Europe where there was still daylight. This provided an interesting weekly opportunity for those city dwellers who spent their weekends in the country. Many major cities, including Prague, are still empty at weekends as people head for their house in the country.

EARLY ATTEMPTS AT HF COORDINATION

It was evident that, as long as jamming continued, any attempt to organize efficient planning of the shortwave broadcast bands would be doomed to failure . Despite this, the 1959 ITU World Radio Conference in Geneva decided that something had to be done at least to improve the chaotic conditions on shortwave. A year later a system had been established where shortwave broadcasters (usually via their country's PTT administration) submitted their schedules to the ITU in Geneva. The data was gathered together and published in the so-called "White Book" or Seasonal Tentative HF Broadcasting Schedules. At first, there seemed to be a ray of hope in this procedure. Despite the limits of international contact as the Cold War progressed in the mid 1960s, I was able to discuss frequency planning ideas with my colleagues from the East Bloc, and also from a couple of Western countries. Yet it was evident that many were losing faith in this project. The more realistically minded among them were already pointing out that the spiral of ever increasing power and number of transmitters would lead us nowhere.

After further quarter of a century of continuing frustrations on shortwave, we are are only a little further nearer the goal. This is despite to large, and expensive ITU World Administrative Conferences convened in the 1980s for the purpose of setting up a system "for the Planning of the HF Bands Allocated to the Broadcasting Service". Each was attended by hundreds of delegates.

NEW EUROPEAN INITIATIVE

It seems that the legacy of the past decades is still very much with us. That is perhaps why a Czechoslovak proposal to set up regular contacts between shortwave frequency managers both from the Eastern and Western part of the continent has taken off so successfully and is already bringing about some measurable results.

The planners met twice in the course of 1991; first in Prague during the spring to amend the frequency schedules for the summer and then again in Sinaia, Romania for the 1991/92 winter period. Further were held for the corresponding periods in 1992. It appears that Austria, Bulgaria, Czechoslovakia, France, Finland, Germany, Great Britain, Hungary, The Netherlands, Poland, Romania, Russia, Sweden and The Vatican are prepared to send their HF frequency planners regularly to these working meetings. The combined complete schedules of this group represent a very substantial part of the global output of HF sound radio broadcasting. The elimination of incompatibilities, even within this group should bring about a significant overall improvement in shortwave reception.

THE FACTS

To understand just how serious the problem facing the planners, it is interesting to look at the statistics which faced us in the early part of last year. During the Prague meeting we processed the schedules of the participating broadcasters. We found a total of 1200 incompatibilities, where stations were either operating on the same frequency at the same time, or 5 kHz either side. About 250 of this total were cases of mutual interference on the same frequency.

Numbers at the meeting in Romania were similar - in fact they were even higher. The increase in interference problems was to be expected since the meeting in August was to consider the northern hemisphere winter schedules. With the longer evenings making propagation on the higher frequencies "close" much earlier, the load on lower frequencies is very heavy.

METHOD OF COORDINATION

A lot of re-use of shortwave frequencies is possible. For instance, under normal propagation conditions, stations in the Pacific can use 49 meter band channels without bothering stations in Europe. The incompatibility problems arise when two - or sometimes more - broadcasters beam their transmissions into the same target area on the same frequency or just 5 kHz either side The target is defined in accordance with an ITU world map, termed as the CIRAF zones of international broadcasting. Each of the 75 zones are divided into four quadrants for a more accurate definition of the intended reception area. The incompatibilities determined by the computer are negotiated between interested parties. Luckily, since schedule information is shared between delegates, consideration can be given to the transmitter power, type and azimuthal bearing of the aerial, and propagation factors as well as monitoring information. Sometimes deletion of an interfering transmission is possible. Alternatively, changes to the azimuthal antenna bearings, or movement by one of the interfering partners to an alternative frequency are the most frequent corrective measures adopted by the coordinators.

Experience has shown that in some cases the frequency incompatibilities detected by the computer program are not too serious when monitored in the target area. Negotiators may then agree that the transmissions "will remain unchanged, subject to monitoring". Naturally, the incompatibility will reappear on the list for the next conference and it will again undergo the scrutiny of the frequency managers involved. It would take much more than the available three or four working days of negotiations to deal with the staggering total of 1200 to 1300 incompatibilities. Consequently, only co-channel cases of mutual interference were dealt with at the two coordination meetings in 1991. In 1992 a meeting in Budapest took the coordination further and the number of direct clashes dropped. The Voice of America joined the coordination group as an observer at the meeting in Evesham, UK in August 1992.

There is a feeling that the reduction in co-channel incompatibilities achieved so far, although small in comparison with the total, nevertheless represents very important first steps in reducing shortwave interference levels. But the continued success of future meetings is also put on a more firm footing thanks to the face-to-face contact between frequency managers. A regularly updated list has been set up with addresses and contact data. The exchange of complete transmission parameters and of monitoring information between participants provides them with much better tools to achieve acceptable solutions to both parties. What is even more important, the process that has been started, may well provide an impetus for a more rational use of the shortwave spectrum. Redundant transmissions can be eliminated, with major cost savings to the participating organizations.

MORE ROOM FOR MANEUVER NEEDED

Promising as this development may be, there is a pressing need to come up with solutions. After decades of intense abuse of the shortwaves, a certain inertia is evident in their present usage. With the East-West tensions now finally over, there is a danger that other urgent measures needed for the restoration of the unique properties and possibilities of shortwave will be simply neglected. After all, the literature is full of the new and attractive means of international radio /and TV/ broadcasting, such as satellites, frequently combined with cable systems, mutual exchanges of local FM facilities etc.

The newly formed shortwave coordination group in Europe can tap into a vast wealth of experience in shortwave frequency planning. All the discussions so far help to underline the urgent need to extend the HF spectrum now available for broadcasting. It has become clear that without more room for maneuver across the entire HF broadcast bands, complete elimination of mutual interference is impossible even amongst a limited group of coordinating countries. The lower part of the spectrum (e.g. 6 MHz) deserves special attention. I completely agree with those of my colleagues who point out that the congestion there was yet another major reason for the failure of all past attempts at coordination.

HOPES AFTER WARC

In addition, there is a recurring - and very annoying - effect which must be tackled. The present spectrum reaches down only as far as 6 MHz band. All European stations operating short distance transmission circuits during the evening and night hours, including Radio Prague International for example, receive complaints from listeners in the winter. The complaints are justified. Even the lowest 6 MHz signals are unable to propagate into the desired nearby areas during such periods. They skip over their target and start to interfere with the reception of other stations within this overloaded band at more distant receiving locations. There was some expansion in the allocation of frequencies to sound radio in the band below 6 MHz at the February 1992 World ITU conference, but not enough to alleviate this problem.

To summarize briefly, efforts by responsible HF frequency planners can play an important role in the present transitional period facing international broadcasting. It is too early to regard satellites as the dominant carrier for broadcasts direct into the home. Shortwave still has a lot of life left in it, so it is important to continue efforts to improve its effectiveness. On a personal note, the opening up of Eastern Europe has meant a great deal to me. At last, professional colleagues on both sides of Europe can communicate and share their experience. I have been gratified to see how quickly colleagues have joined in the spirit of cooperation. Special help in the preparation of documents for the meetings and in organizational matters has been provided by the Deutsche Welle, namely by my colleague Horst Scholz.

We can recall that the Cold War in radio started here in Central Europe about 40 years ago. It is therefore logical that the initiative to do away with its legacy on shortwave now comes from the same region.

ABOUT THE AUTHOR

Oldrich Cip, better known to international shortwave listeners as Peter Skála, is editor of the DX column at Radio Prague International and founder of the Monitor Club . He holds a degree in linguistics but most of his more than 30 years in radio has been spent as a shortwave frequency planner and broadcast coverage engineer.

The political suppression of the Prague reforms took place in 1968. Because of his views and contacts with Western broadcasters, Oldrich was banned from travelling abroad and was forced to stop his attempts at international cooperation in shortwave radio. Early in 1990, after the success of the Czechoslovak "Velvet Revolution", he suggested that a new East-West initiative in shortwave frequency coordination should be started in Europe. Oldrich is married, has two teenage sons and lives in Prague.

CHAPTER 8
RECEIVER TEST RESULTS

DAK MR-101

In mid 1990, a small portable receiver appeared in the US. It was manufactured in the Peoples' Republic of China for a large mail-order house based in California. Called the DAK MR-101, it turns out to be quite a compact travel portable.....18 cm by 12 cm by 4 cm. It works off four penlight cells which give around 20 hours of continuous use. The price caused a lot of discussion in North American shortwave circles...just under US$50 without the shipping charges. Models sold in 1990 offered a radio with medium and shortwave capability, plus FM in mono, PLL synthesized tuning, and a digital frequency readout display that even had a light in it to view it in the dark. On the face of it, the MR-101 has features that compare well with a radio costing five times the price. In the course of 1991 the radio was improved with the addition of FM stereo and some minor improvement to the receiver's image rejection. An ON-OFF timer option was also incorporated into the newer version. The price was kept the same but the type number changed to MR-101s.

So what do you get? The set has four frequency bands. mediumwave is between 530 and 1630 kHz, but it tunes in 10 kHz steps. You can't switch that to the 9 kHz spacing that stations use if you're travelling outside North America. In Europe for instance, the Dutch domestic service channel on 675 kHz can't be tuned in successfully at all because it's midway between the steps. On shortwave, you can tune the set in 5 kHz steps, but you can't fine tune in between that. Coverage is between 3.2 - 7.3 and 9.5 - 21.75 MHz which means that most of the out of band channels used by international broadcasters are covered. But useful channels like 9410 kHz from the BBC cannot be received. The radio has a total of 20 presets, five per band.

Tuning is by means of two buttons, "up" to increase the frequency, and "down" to do the opposite. It takes a bit of getting used to because if you hold down either button for a few seconds the radio switches from tuning at a normal rate to a very fast rate indeed. That can mean you overshoot the frequency you were looking for. Picking a channel is made easier by the digital frequency readout, though if the MR-101 shows 11755 for instance, the last 5 is much smaller on the display than the other figures. That might lead to a bit of confusion at first. The MR-101s changed the last digit so that it is the same size as the others.

The radio's sensitivity is fine for Western Europe, although in lower signal strength areas you might have to extend the telescopic whip fully (or connected an external antenna) to pull in the international broadcasters. The selectivity is fair for a radio of this type. The biggest problem is image rejection. If you try to listen to Radio Moscow on 12030 kHz and then punch in a frequency exactly 900 kHz lower, in this case 11130 kHz, you get the same station with almost equal strength.

These internally generated signals are a nuisance on the 49 meter band because amateur radio operators using the 40 meter ham band (around 7 MHz) break through and spoil many broadcast signals on the 49 meter band, 6 MHz. It is strange to operate a radio that is full of broadcast signals from 5 MHz right up to 14 MHz without a break - half of the signals shouldn't be there.

The radio has a built-in clock and alarm which is handy for travelling. It comes complete with a 12 page guide to operating it, which for the most part is clear, and a frequency guide which is several years out of date. The frequency guide has been taken directly from a competing radio company in Taiwan, who in turn based theirs on a Japanese company's "Wave Handbook". It is nostalgic to look back at schedules from 1983, but not very practical.

But what do you expect for 50 dollars? In the US market it is clearly a useful product to get people interested in shortwave listening, and it is perfectly adequate for listening to the stronger international broadcasters.

Outside the region it is less successful because of the non switchable 10 kHz tuning on mediumwave, and the inability to handle strong shortwave signals, like you'd find here in Europe. But it's an encouraging trend. Drew Alan Kaplan, who's founded DAK industries 26 years ago is clearly a keen international radio enthusiast. He says that to broaden the market for international radio, the price is certainly part of the key. However, at presstime (October 1992) DAK industries had filed for bankruptcy protection under the US "Chapter 11" scheme, but they were still trading. More information from DAK Industries Inc, 8200 Remmett Ave, Canoga Park, CA 91304 USA. Tel: (toll-free) +1-800-888-7808, or +1 818 888 8220

DAK DMR-3000

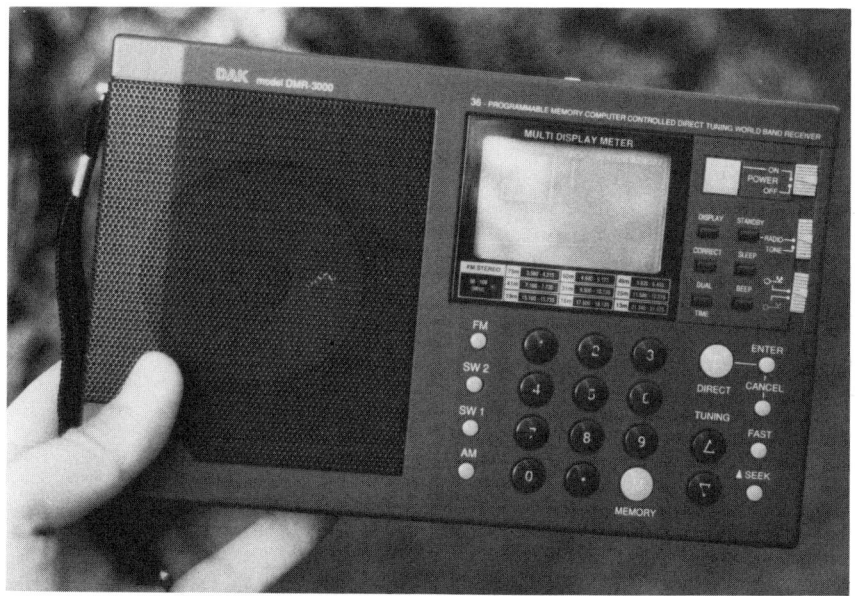

In May 1992 DAK Industries launched a more up-market shortwave portable which offers better results than the MR-101s. The radio measures 21 x 13 x 5 cm and weighs in at 814 grams including the 4 penlight (AA) batteries. When switched off the radio liquid crystal display is blank. You have to switch the radio on and select the clock to be able to see the time. The time in a second time zone can also be displayed. Pushing the CLOCK/RADIO button activates the radio and the time disappears. The frequency is show in MHz. On shortwave the readout is in line with the channel spacing, i.e. 5 kHz. If you are tuned to 11835 and move up by pushing the UP button, the radio jumps to 11840 kHz. No fine tuning within that 5 kHz is possible, and that is a serious disadvantage. FM stereo is activated by plugging in a pair of headphones, otherwise reception is in MONO on the built-in speaker. A switch in the battery compartment allows two types of tuning on AM (530-1710 kHz in 10 kHz steps, or 522 - 1620 kHz when set to 9 kHz).

The radio offers four tuning "ranges": mediumwave AM, as described above, FM (87.5-108 MHz), and two shortwave ranges, 3580-7735 and 9500-21975 kHz. This means that some important out of band frequencies are missing. Note too that coverage of the shortwave bands is NOT continuous. For instance, if you tune up from the 31 meter band you notice the radio jumps from 10135 kHz to 11580 kHz, and you cannot get channels in between. That said, coverage of the broadcast bands is quite generous, only some very out of band channels from Beijing are not included, plus the BBC's popular 9410 kHz. The new 15 meter broadcast band is also not tunable. Each of the bands (AM, FM, SW1 and SW2) offers 9 preset memories under the keypad numbers 1-9. You just press the key to recall the memory. Direct entry tuning is more complex. You have to select a direct entry key, tap in the frequency, and then touch a tiny button marked "Enter". People with a physical or visual disability may find that a bit inconvenient. There is also a SEEK function where the radio scans up the dial stopping when it reaches the next signal on the air. The dial can be illuminated with a light, and a single step attenuator switches in around 25 dB of attenuation for strong local signals. In practice that is useful for mediumwave listening, but its use on shortwave makes the set totally deaf.

The built-in speaker gives fair fidelity, though it is not as rich as more expensive receivers of this size. We found the volume control on the side of the set was very easy to knock and started crackling after a few months heavy use. The white numbering also came off the keyboard.

In short, at a price of just under 70 US dollars this receiver doesn't offer all that more than the MR-101s. The performance is slightly better, although image rejection is still only fair. For a few dollars more you could get a good analogue portable with far more tuning versatility. The ability for "fine tuning" on the shortwave broadcast bands is important. More information from DAK Industries Inc, 8200 Remmett Ave, Canoga Park, CA 91304 USA. Tel: (toll-free) +1-800-888-7808, or +1 818 888 8220

DRAKE R8

The arrival of the R8 in mid-1991 was accompanied by a major advertising campaign in both Europe & North America. After all, Drake made a popular high-end communications receiver in the mid-80's and then departed from the shortwave market to concentrate on satellite receivers. Now that they're back, the influence from the satellite business is evident in the styling of the receiver. It is compact, and at 5.9 kg, surprisingly light. Internally the receiver is sturdily made, and easy to service. Despite what you see in some advertisement photos, it is easiest to use when tilted at a slight angle towards you. Otherwise punching in the frequencies on the keypad is rather awkward. The keys have a "rubbery" feel, and are quite close together.

When you're designing a radio you have two options. You can put a button on the front panel for every function. However, that can lead to a radio looking somewhat like the cockpit of a jumbo jet. On the other hand, you can reduce the number of buttons by putting a number of functions under one key. This latter approach has been adopted by Drake with their R8. This simplifies the panel, and can bring down the cost. But it can make the receiver awkward to use.

Take the mode button for instance. You can chose between AM, narrow band FM, CW, RTTY, lower-sideband, or upper sideband, in that order. But if you're listening to a station in upper sideband, and you want to check the lower sideband, then you have to press the mode button five times. That's a nuisance. Likewise, the default bandwidth selection is related to mode as follows:

Mode	Default Bandwidth (kHz)
AM	6
FM	12
CW	1.8
RTTY	1.8
USB/LSB	2.3

If you select a station in USB (2.3 kHz default) and decide that listening with a slightly wider filter would be more enjoyable, then you have to cycle through all the bandwidth options to reach 4 kHz.

TUNING STEPS

The tuning steps are also related to mode. If you select the AM mode, the tuning steps are in 100 Hz, and the display shows the frequency to the nearest 1 kHz. You make this finer by a factor of ten if you want, so the display shows frequencies to the nearest 100 Hz. However, on the first models if you then touched the mode switch, the radio switched back to its 1 kHz default setting. This problem has now been solved. In the overcrowded shortwave broadcast bands, use of the 4 and 2.3 kHz filters is more difficult if the tuning steps are 100 Hz. If you listen on SSB however, then the resolution is to the nearest 10 Hz, and the tuning steps are 10 Hz too. Drake say they've done this deliberately to make it easier for the user. Our practical experience shows that, e.g., having decided to listen to a broadcast station in LSB with a 4 kHz filter, it is extremely annoying to have the receiver go back to a 6 kHz filter and 1 kHz display resolution when you decide to check the AM mode.

You get a set with a wide choice of bandwidth filters with this receiver, and the shape factor of these filters is fine. You get a choice of 6.4, 2.3, 1.8 & 0.5 kHz. But if you keep changing the filter settings to reduce interference, you must remember to retune.

PASSBAND COMPLICATIONS

Normally, when you switch from a wider bandwidth filter to a narrower one you expect the audio quality to drop. The interference from nearby stations should also be reduced, and at the same time the sensitivity of receiver should increase. When we switched to a narrower bandwidth filter on the R8 we found that the sensitivity went down instead, almost by a factor of three. Of course there was a reason....the adjustment of the "passband tuning control" is critical. We have since checked other examples of the R8 in the United States to find that the factory setting of the passband does vary from example to example. In our production model, to get the best results every time you switch bandwidths, you needed to keep adjusting the bandpass control for maximum intelligibility. That took some getting used to. Note that the pass-band tuning cannot be switched out of circuit.

Coverage on this set starts at 100 kHz, so in Europe if you're interested in utility signals below that, you're out of luck. In practice there's such a strong oscillator whistle at 100 kHz that coverage really starts at 110 kHz. There is an RS-232 connector on the back of the receiver for computer control. Drake have launched their own software (IBM PC Compatible). This allows extended timer functions for multi-event recordings as well as up to 1000 memories and VFO capabilities for rapid movement from one channel to the next. The operating parameters can be stored on (hard)disk. The software package costs a very reasonable US$60. Alternatively the program "SW Navigator" from Jim Frimmel is able to control the receiver using a Macintosh computer, and adds a lot of extra tuning functions too plus a frequency database (cost US$99). The development of such software is not as easy as on rival receivers (e.g. JRC NRD-535, Kenwood R-5000, Icom ICR-71).

DIGITAL DISPLAY

When it comes to the display, Drake has decided to make the R8 show the carrier frequency that's tuned in. When you're listening to broadcast stations of course, that's exactly what you want. But if you tune in a Morse code station to exactly the frequency of the carrier wave you'll hear nothing at all. So you have to detune the signal around 850 Hz lower in order to get a tone. 850 Hz is a standard used in Europe. You need to keep this in mind when referring to frequency lists.

The same is true of RTTY signals. Depending on the type of decoder in use, you'll have to tune the set some 1375 Hz higher than the center frequency to be able to read the text being sent over the air. Many other shortwave communications receivers give you the option of switching in that compensation automatically. In this case though, if you're a utility enthusiast, you'll have to remember to detune the set by 1375 Hz from the frequencies you'll find listed in publications by Klingenfuss, Grove, or Gilfer Associates. We don't think that's too handy.

SENSITIVITY

The average sensitivity for 10 dB s+n/n in microvolts was measured at 50 Ω.

Freq Range in kHz	AM 6	AM 4	AM 2.3	SSB 2.3	SSB 1.8	CW 0.5
100-470	2.60	1.30	1.00	0.70	0.70	0.31
470-1800	1.40	0.8	0.7	0.36	0.35	0.16
1800-30000	1.70	0.88	0.75	0.37	0.36	0.15

Like most sets in the same price range, you don't need to worry about the R8's sensitivity. We measured figures of around 1.4 microvolts using the 6 kHz wide filter for shortwave broadcast listening, and just a little less sensitive on mediumwave. These are good figures. The preamplifier only works above 5 MHz, multiplying the sensitivity by two. The attenuator gives 10 dB damping. The RF-gain control is rather lop-sided. For the first half-turn nothing happens, whilst the second half knocks back the sensitivity from 0 to -120 dB. The "S"-meter is more of a tuning indicator. Below S5 the signal is shown as slightly below the real value, above S5 the meter shows the signal to be a bit stronger than it really is.

SYNCHRO-DETECTOR

There are two ways to build a synchronous detector. The first is that the receiver locks onto the carrier of the broadcast signal, and the user then selects either the upper or lower sideband mode using a narrow filter. If the radio remains locked, music and speech sound at their natural pitch. Without the lock, music in particular sounds out of key.

The second way to design a synchronous detector is to let the radio lock onto the carrier, but the radio stays in the AM mode. The passband of the receiver can then be adjusted (either higher or lower in frequency) so that interference from stations either higher or lower in frequency can be reduced, or even eliminated. Adjustment of the passband control is critical....if you just press the sync button, the effect is hardly noticeable.

Theory says that by selecting to listen to only one sideband, the audio quality is improved, especially during "selective fading". Distant signals, arriving at a low angle with respect to the horizon, often exhibit phase differences between the upper and lower sidebands. This can lead to distortion in a standard AM detector as one sideband partially cancels out another. Since the sidebands carry identical audio information, only one is really needed. In practice the percentage of shortwave signals with noticeable selective fading is quite low. The main advantage of the SYNC option is the ability to escape interference affecting one sideband, and, at the same time, retain the correct pitch on music and speech.

The Drake R8 has a built-in synchronous detector. It's marked as a single push button on the front panel. First you must let the set lock onto the carrier of the distant signal. That can take a few seconds. Having done that, you can then adjust the passband tuning control, Because the set has now locked onto the carrier, you won't notice any change in pitch....which is what you want. We conclude that the sync on the Drake R8 works well. There is a noticeable drop in the distortion level when you use the synchronous option (from 3 to 1%). The set locks well onto even the weakest signals, e.g. using the 6 kHz filter the sync locked onto a signal of just 0.35 microvolts. This represents a signal only just above the noise floor of the receiver!

NOTCH FILTER

The notch filter in this set is actually an audio notch filter. That means that intermediate frequency stages and the automatic gain control (AGC) can still be influenced by a station very nearby in frequency. This means that the audio from the interfering station may be reduced, but the AGC is still knocked back. This results in distortion on the weak signal you're trying to hear. Other manufacturers (e.g. Kenwood & JRC) put the notch in the intermediate frequency stage of the receiver, which has a number of advantages. Then it is possible to attenuate heterodyne whistles, so that the receiver's AGC then reacts to the wanted signal, and the audio from the unwanted signal is effectively removed. In our tests, the range of the audio notch is 520 - 4200 Hz. Drake claims the notch attenuates by some 40 dB (i.e. 100 times) but that proved impossible to reproduce in practice. At 520 Hz the attenuation was just 16 dB, and at 4200 Hz the figure proved to be 33 dB.

The automatic gain control is designed to iron-out some of the audio level fluctuations caused by shortwave fading. Drake have designed an excellent AGC, with an attack time of 1.3 milliseconds. This is ideal for band scanning.

DYNAMIC SELECTIVITY

Drake says that the dynamic range is better than 90 dB. That is clearly measured with the 500 Hz filter, which would normally be useless for listening to shortwave broadcast stations. If you listen to AM broadcast signals using the 6 kHz filter, then the dynamic range is 79 dB. In the SSB mode, using the 2.3 kHz filter, our measurements indicate a dynamic range of 84 dB. This is a good figure, slightly worse than the R-5000 from Kenwood. The 3rd order intercept point, using SSB, works out to be -0 dBm, and +3.5 dBm using the CW filter.

The tone control is somewhat primitive. It simply boosts or cuts the bass by 6 dB (at 200 Hz), and we thought it could have done more. Birdies...there are quite a lot for a receiver of this type, but they are very weak. Birdies above one microvolt are noted on 15970, 17970, 19970, 20370, 20770, 21165, 21965 and 27160 kHz.

The set is incredibly stable, to with 1 Hz per half-hour. You also have access to a 100 channel memory and clever scan facilities.

CONCLUSIONS

In late 1991 Drake issued a new software PROM which solved some of the critical points on our test model. The annoying factory-set "defaults" linking the frequency display/tuning rate and the AGC response have been eliminated. The user can now select important criteria corresponding to each mode, and then the radio defaults to these setting each time that mode is selected. You can now get the radio to display frequency readout to the nearest 10, 100 or 1000 Hz, regardless of the mode. Existing owners of the first production run can upgrade by contacting their dealer for a PROM upgrade, retailing at about US$14.00, excluding postage. The software revision doesn't make it easier to select bandwidths or modes, without cycling through all the possibilities each time.

In the United States, the Drake R8 costs around US$960. Optional VHF in-built converters cost an extra US$190. In Germany the set is advertised as the "follow-up to the famous Drake R7". We don't think the two sets are in the same league. That said, the R8 is a very complete and pleasant sounding package, ideally suited to people who enjoy serious program listening. Whilst it is clear than Drake is willing to listen to critical comments, there are still annoying quirks in the R8 which take some getting used to.

It would not be surprising if this set is in fact the forerunner of other new receivers..... there's a market for a knock-down version of the R8 (without sync for instance), and many people would welcome a more up-market receiver to compete with products from JRC and the Icom R-9000.

In Europe the price of the Drake R8 is in the region of 3200 German Marks or £965 sterling in the UK. In Germany that makes the R8 just 400 Marks short of the JRC NRD-535. That in our opinion, makes the set far too expensive. More information from R.L. Drake Company, P.O. Box 3006, Miamisburg, Ohio USA. Tel: +1 513 866 2421. Fax: +1 513 8660806. Drake Canada, 655 The Queensway #16, Peterborough, Ontario, Canada. Tel: +1 705 742 3122. Drake Europe, Dr Trueta, 1 y 3 entresuedo, 08860 Castelldefels, Barcelona, Spain. Tel: +34 3 636 0192.

GROVE SW-100

Originally scheduled for launch in August 1992 this receiver was not available for purchase when we went to press. A hand built example of the Grove SW-100 was on show at the Dayton Hamvention in April of 1992. If the specifications sheet is anything to go by the radio should be an interesting receiver going head to head with the Drake R8, Lowe HF-250 and Yaesu FRG-100. But all we have at the moment is a photograph.

GRUNDIG YACHT BOY 206

This receiver has been designed by Grundig, although this appears to be the first time the German company has gone to a Taiwanese company to have it manufactured. Product Manager for Audio Manfred Lichius explained to us that they want to keep a low-cost analogue portable in their line-up, despite the fact that digital portables with a synthesizer can now be made for the same price. Analogue sets still offer a much lower background noise, and generally better performance.

The Yacht Boy 204 is identical to the 206, except that the more expensive 206 has a digital clock and timer built-in. You can also set the radio to switch off after a preset period, which is useful if you want to hear music to send you to sleep.

The 206 offers no less than 15 bands, a lot more coverage than many sets in the same price bracket. Medium & longwave are available, plus FM in mono. shortwave coverage of the 120, 90, 75, 60, 49, 41, 31, 25, 22, 19, 16, and 13 meter bands is provided, including plenty of coverage either side of the "official" bands. The set weighs 555 grams, including the three AA-cells needed to power the radio, and the single AA battery used for the clock/timer. A plastic stand on the back of the set allows the radio to be tilted when operated on a table. If the stand is accidentally hit (e.g. by accidentally dropping the set) it is designed to pop out without snapping off. Replacement is easy.

Power consumption is remarkably low, averaging around 25 mA at a comfortable listening volume, even on FM. Band spread on the receiver is good, the tuning knob having just a slight backlash. Despite the single-conversion design, the image rejection is quite good for a receiver of this type. Listening in the evening hours on the 49 meter band there were surprisingly few "ghost" signals.

At a price of around US$70 in Europe, this set offers excellent value for the traveler looking for a no nonsense shortwave radio.

In North America the set is more expensive, retailing at around US$120. Since the Sony ICF-7601 is also retailing at around the same price a comparison of these two is worthwhile. The Sony gives better shortwave performance (thanks partly to its dual conversion design) but you may prefer the fuller audio of the Grundig. More information from Grundig, Kurgartenstrasse 37 Fuerth/Bayern D8510, Germany. Tel: +49 911 7030. In the USA, Grundig has a distributor. This is Lextronix Inc, 3520 Haven Ave. Unit L , Redwood City CA 94063, USA. Tel: +1 415 361 1611 or in the USA only call (800) 877 2228. In Canada call (800) 637 1648.

GRUNDIG YACHT BOY 220

This set is assembled in Taiwan, but has been made to Grundig's own specification (i.e. it is not a standard set from another company with a Grundig badge stuck to it.). The receiver is priced at DM180 (US$99) in Germany, putting it alongside the Sony ICF-SW20 in terms of price. Contrast this to the United States where the Yacht Boy is considerably more expensive (around US$150) whereas the Sony ICF-SW20 is around US$90.

The Yacht Boy 220 is larger than Sony's compact SW1, though at 14 x 3.5 x 9 cm it is not much larger. It does not fit into a shirt pocket, but it is small enough to fit into a suitcase...a simulation leather pouch is provided with the set to protect it. The Yacht Boy 220 weighs 400 grams including the three penlight batteries. An optional 4.5 volt DC mains converter is available, although since one set of batteries gives around 35 hours continuous use, you would probably take two sets of batteries on holiday rather than carry a transformer pack and a plug adapter.

COVERAGE

The Yacht Boy 220 does not have continuous shortwave coverage, but all the international broadcast bands are included (ref: Sony ICF-SW20). The model we examined had the medium and longwave bands, FM from 87.5 - 108 MHz (in mono), and nine shortwave bands, i.e. 75, 49, 41, 31, 25, 21, 19, 16 and 13 meter bands. However, Grundig has not been quite as generous as Sony in its coverage of out of band frequencies. On 31 meters for instance, BBC on 9410 kHz or Tehran on 9022 kHz were just off scale. BBC on 15070 kHz was tunable, although it was right at the edge of the dial.

This receiver has been designed to give maximum coverage for the price charged. It is a broadcast only receiver (i.e. no SSB) and the pointer-and-dial analogue readout allows accurate frequency readout to within about 20 kHz. The rotary tuning knob had a slight backlash on the model we tested. The accuracy of the readout was fair...on the 19 meter band the scale is not quite linear.

The 220 is a straightforward design with a single AM bandwidth filter and only single-conversion circuitry. This is sharp enough to separate stations 5 kHz apart with gentle fine tuning. The radio is more selective than the Sony ICF-SW20. There are three tone settings, for speech, normal and music. This seems rather limited for a set of the late eighties...it seems strange that this is cheaper than a variable control. We found the "speech" setting to be the best when using shortwave, although this is probably a matter of personal taste. As you turn down the sound, the bass response is increased (a kind of loudness control mainly noticeable on FM). Audio quality on FM was crisp and rich, despite the radio's small internal speaker.

Sensitivity for a radio of this type was more than adequate. The radio has been deliberately "de-sensitized" on mediumwave to avoid overloading problems on the European AM dial. You would not use such a set for mediumwave DXing (i.e. difficult long distance listening) anyway. Sensitivity on shortwave is fine using the built-in telescopic antenna. Retracting the antenna is needed during the evening hours to avoid overloading the front end circuitry. An external antenna is not advised, except may be in very remote low signal- strength locations.

CONCLUSIONS

Although Grundig has not been so generous with out of band frequency coverage as the new Sony ICF-SW20, the Yacht Boy 220 has more meter bands including the important 21 meter band! In Europe it is clearly designed to compete with the Sony. The Grundig Yacht Boy 220 offers better value/performance in Europe. In the United States, the retail price is US$119, though you can expect larger dealers to discount this down to the region of US$99. More information from Grundig, Kurgartenstrasse 37 Fuerth/Bayern D-8510, Germany. Tel: +49 911 7030. In the USA, Grundig has a distributor. This is Lextronix Inc, 3520 Haven Ave. Unit L , Redwood City CA 94063, USA. Tel: +1 415 361 1611 or in the USA only call (800) 877 2228. In Canada call (800) 637 1648.

GRUNDIG SATELLIT 500

Despite being announced and previewed in January 89, stocks of this radio didn't reach the shops in Germany until the end of March 89. There were initial problems with quality control in the first batch of receivers (mainly complaints about overloading) but these were quickly improved by the manufacturer.

Grundig's distribution on the home market in Germany, and more recently in North America, is far better organized than in other parts of the world. The Grundig Satellit 500 sells for 750 German Marks in Germany, though at US$499 (discounted price) in North America it is considerably more expensive on the other side of the Atlantic. The Satellit 500 is now being discounted at many outlets (mid 1992) to make way for the Satellit 700. Let us examine the features you get on the Satellit 500.

TRANSPORTABLE

The Satellit 500 is a portable receiver designed for use in the home, or perhaps in a mobile home or caravan. It weighs 1800g with the batteries inside, so you would probably choose a different radio if you wanted something simple to take on a plane with you. A piece of A 4 size paper just covers the front panel, and the set is about as thick as a 900 page reference book, - something like an English dictionary. There is a handle on the top of the black plastic case so it is easy to carry around. The set has rounded sides, so there are no sharp edges on the case. For the record, the dimensions are 300 x 180 x 70 mm.

With the exception of two small slide switches on the side of the set, the radio is controlled by push buttons and rotary controls on the front panel. Half of the front of the radio is taken up by the loudspeaker. On FM, this speaker is capable of producing rich treble and bass for which Grundig is famous. The right side of the radio contains a keypad for tuning the set, and a liquid crystal display to see what you've tuned. If the receiver is connected to a mains power unit, the display and the tuning section of the keypad are permanently illuminated (even when the set is on standby).

LIGHT NOT ACTIVATED BY MANUAL TUNING!

When you are using batteries, touching any of the keys activates the light for a period of 10 seconds. That is handy if you wanted to tune the set at night without switching on a light in the room. However, if you first key in a frequency and then start using the manual tuning knob, the light still cuts out after 10 seconds, leaving you literally in the dark.

THREE VERSIONS FOR THREE MARKETS

The Grundig Satellit comes in three versions. In the English speaking parts of the globe, the international version will be the one seen. It covers the longwave band from 148 to 353 kHz, and the entire mediumwave and shortwave bands continuously from 513 kHz right up to 30,000 kHz. FM is included from 87.5 to 108 MHz as well, so this is a full coverage radio. There is a version sold in Germany, and another one in Italy which has reduced shortwave coverage because of local legislation.

TWO CLOCKS

When switched off, the set shows one of two programmable times. There are two linked clocks in the set, so you could set one to local time, and the other to UTC. A key on the front of the radio brings the radio to life, though there is a lock facility to prevent this happening if the set was put in a suitcase. When switched on, the radio selects the last frequency it was tuned to.

There are three ways of manually tuning the set. There is a familiar notched tuning knob which rotates in noticeable "steps". A visually handicapped person would be able to count the steps and know where he or she had tuned to from a known reference point. The radio moves in 1 kHz steps on standard AM, 100 Hz in the case of SSB or in the synchronous detection mode, and 25 kHz steps on FM. These steps are small enough for AM broadcast reception. We found them too coarse for radio teletype reception or easy demodulation of radio amateurs using single sideband. Steps of 20 or even 10 Hz increments would have been better. As you turn the tuning knob, you can hear the radio move in audible steps.

KEYBOARD ENTRY...SMART SOFTWARE

You can select a frequency directly using the keyboard, and here Grundig have spent time ensuring the software is intelligent. You can type in 6.165 or just 6165 to select that channel on shortwave, It does not matter whether you think in Megahertz or kHz. If you think in meter bands, then any number typed into the radio below 100 is interpreted as meaning a meter band. If you tap in 49 for instance, the Grundig Satellit 500 jumps to 6075 kHz, a Deutsche Welle frequency in the middle of the 49 meter band.

The radio has a search facility, allowing you to scan within a chosen broadcast band in 5 kHz steps. That is handy if you are just curious as to which stations are around. If you find something you like, there are up to 42 free positions in the radio's memory banks. In addition, you can add up to 4 alphanumeric symbols to remind you what the station was...so you might choose "VOA1" to stand for a Voice of America frequency and so on.

The Grundig Professional Satellit 500 sold only in Germany has a number of station frequencies already programmed into the radio. Although it sounds a nice idea, the stations chosen would probably not match the interests of an English speaking listener, and there is no way the radio can keep up with the constant changes happening on shortwave. Radio Netherlands, for instance, is given two shortwave channels no longer in use for Europe, and 747 kHz is erroneously allocated to the Dutch external services. It might have been a better idea to program the radio with known mediumwave channels which don't vary from season to season. Frankly, we found the pre-programmed stations facility to be more of a gimmick. There is a way to convert an "International" version of the receiver so that the pre-programmed frequencies are activated as in the Professional version. However, Grundig discourage the practice, and say this will invalidate the warranty.

COMPARISON WITH SONY ICF-2001D

Grundig has been making shortwave receivers for 25 years, and from its features, it is clear the Grundig Satellit 500 is designed to compete with the slightly smaller Sony ICF-2001D (or ICF-2010 as it is known in North America). Sony's radio, launched in 1985, is US$50 more expensive in Europe, yet in the US where the 2010 retails around US$360, the Sony works out US$40 cheaper.

The Sony 2001D made its name in the shortwave community because of what's termed synchronous detection capability. Although not a new concept, this was the first set to bring such a facility within the price range of the serious shortwave listener. Synchronous detection is difficult to explain in non-technical terms, so we will have to oversimplify a bit

SYNCHRONOUS DETECTION

Basically a short or mediumwave radio signal consists of two components. There is a carrier which puts the station at a certain point on the dial, and two identical sidebands which contain the music and speech you are trying to hear. It is important that the strength of the carrier is in about the right ratio to the level of the sidebands, or you get terrific audio distortion on a simple receiver. But on shortwave, received carrier levels are going up and down because of the imperfections of the ionosphere. You can see this on the signal strength meter. Suppose that you set up a very stable oscillating circuit inside the radio, often called a beat frequency oscillator. This could be set to operate in exactly the same phase as the incoming carrier, and help to iron out the variations due to the ionosphere. Then you can continue to correctly demodulate the sidebands, even though the carrier may be fading in and out all over the place.

On many communications receivers you will find a single sideband facility. You can switch this on, tune into a broadcast station, and adjust the tuning knob very carefully until music and speech sound normal. That is not quite the same as synchronous detection on the Sony or the new Grundig radio. In the sync mode, both sets lock onto the incoming carrier automatically, using it as a "pilot" tone, and making fine tuning adjustments unnecessary.

STABILITY

For SSB and RTTY reception, stability is important. At a constant room temperature of 18 Celsius, the radio drifted less than 10 Hz in one hour. This is excellent. However, the radio is not easy to use for radio-teletype reception because the smallest incremental tuning step in 100 Hz.

NO SELECTABLE SIDEBANDS ON SYNC MODE !

On the Sony ICF-2001D you can decide whether you want to receive either the upper or lower sideband in the sync mode. On the Grundig you cannot do that, which we think is a real shame. However, it must be said that the selectivity of the Satellit 500 is considerably better than the Sony. The bandwidth filters offer static selectivities of 5.65/18 kHz and 3.7/9.6 kHz measured at -6/-60 dB. Further measurements show the curves are sharp. You will find less annoying whistles from nearby stations on the Grundig, although the audio quality on shortwave therefore lacks some of the higher frequencies. Whether you describe the Grundig's AM audio as mellow or muddy depends very much on personal taste.

SENSITIVITY

The four input filters, which switch automatically depending on the selected frequency, affect the receiver's sensitivity across the frequency range covered. We could detect no difference between sensitivity in AM normal and AM SYNC positions, but, as you would expect, a slight difference (6%) in sensitivity depending on whether AM WIDE or AM NARROW was selected. The following results were achieved:

Sensitivity in microvolts for 10 dB S+N/N (AM NARROW, 1 kHz, 60% mod)

MHz	AM	SSB
1.6 - 4	2.3	0.9
4-8	3.2	1.25
8-20.5	1.2	0.47
20.5 - 30	1.1	0.43

The Satellit 500 has an internal ferrite rod antenna for long and mediumwave reception which cannot be switched out of circuit. For FM and shortwave, the telescopic whip is used. Unlike its predecessor, if you fully extend the aerial (115 cm), the radio does not fall over. In fact for shortwave, the whip is in fact a compact active antenna. You can plug in an external antenna if you wish, and switch off the internal one so they do not try to compete. Grundig has gone to considerable trouble to ensure that overloading of the delicate front end circuitry of this receiver by strong short or mediumwave signals is minimal. This is done by automatically adjusting the input circuitry depending on the frequency the receiver is tuned to, even on medium and longwave!

RF GAIN CONTROL

Turning the RF gain control out of its locked position switches off the radio's automatic gain control. You can adjust the set's sensitivity over an enormous range (90 dB). In addition there is an attenuator which gives an extra 20.1 dB reduction in signal strength if needed. The AGC action on this radio is well designed. Once the receiver gets enough signal to reach the intelligibility level (S/N 10 dB) then the volume remains at a constant level despite fluctuations in signal strength.

The blocking level for the Satellit 500 is reasonable. It is possible that strong signals within 100 kHz will cause reception problems to the channel you are listening to, but in practice this would only occur in very high signal strength conditions.

Blocking figures for Satellit 500

Freq. Distance in MHz	Signal level in mV	dB
±0.1	7.3	62
±0.2	8.2	63
±1.0	11.6	66

Dynamic Range

	AGC ON	AGC OFF
Dynamic Range	66 dB	72 dB
3rd order IP	-12 dBm	-1 dBm

Although switching off the AGC results in a slight loss of sensitivity, the dynamic range improves. Under high signal strength conditions, turning the RF gain down will make little difference to the signal strength (any drop in volume can be adjusted with the volume control) whilst the intermodulation products drop back considerably.

We feel that Grundig has got selectivity, sensitivity and dynamic range about right for a set of this type.

ECONOMICAL BATTERY CONSUMPTION

We ran the set continuously on a fresh set of alkaline batteries and got around 108 hours of life before noticeable distortion set in. The Satellit 500 is more economical in its battery consumption than the Sony ICF-2001D.

Reception on FM is possible in stereo by plugging in an optional speaker, or a pair of stereo headphones. You can also use two built-in times to remotely switch a tape-recorder on and off via a relay unit. The radio will not switch the mains voltage.

BIRDIES

All receivers in this price range suffer from internally generated whistles. If these coincide with a desired frequency, then reception can be impaired at best, or totally impossible. We put the Satellit 500 in a Faraday cage, selected the USB mode, and tuned slowly between 1612 and 30 MHz. Only birdies with a strength greater than 6 millivolt (S/N >20 dB) were noted. Overall the radio is fairly quiet, except in the range 5752 - 6538 kHz. We found strong birdies on 5937.1, 6012.4, and 6275.0 kHz. Other strong internally generated signals were found on 10485, 12583, 15737, 16770, 18825, 20971, & 25165 kHz. Weaker birdies were found elsewhere, but they are unlikely to greatly affect reception

POOR INSTRUCTION BOOKLETS

The radio comes with two booklets. One is simply a technical guide to shortwave. Sadly this has such a complicated style that its likely to put off the beginner for life. The instruction booklet has been literally translated from the German and contains a number of clumsy phrases. Even the owners manual can't decide whether to call the radio a "Satellit" or a "Satellite". How many users will recognize that MEZ stands for "Mittel EuropÑische Zeit", or Central European Time or that "HA" in the radio's display stands for Handabstimmung (which means "manual tuning")? Grundig in North America once gave away a videocassette introducing the new radio with the Satellit's more expensive cousin, the 650. It would have been nice if they had done the same with this one.

SUMMARY

Here is a solid, well-built portable radio set. It has good performance. Some of the tuning functions are rather complex, and unless you use things like scanning on a regularly basis, you will have to keep consulting the manual to remember the keying sequence to use. If you are tuned to an SSB signal, and decide to punch in a new frequency (rather than using the manual tuning knob), the radio jumps into the AM mode, and you have to reselect USB or LSB. The set is capable of producing quite a high volume, and the overall audio is more mellow than equivalent Japanese competition.

Synchronous detection is available, but its effectiveness is handicapped by not being able to select either the upper or lower sideband. In Europe, the Satellit 500 has already disappeared from the market. Even though the set has been recently discounted in North America (now around US$399) check the price of the competing Sony ICF-2001D (around US$360 although production discontinued mid 1992). Performance wise, the Sony ICF-2001D offers better value.

More information from Grundig, Kurgartenstrasse 37 Fuerth/Bayern D8510, Germany. Tel: +49 911 7030. In the USA, Grundig has a distributor. This is Lextronix Inc, 3520 Haven Ave. Unit L , Redwood City CA 94063, USA. Tel: +1 415 361 1611 or in the USA only call (800) 877 2228. In Canada call (800) 637 1648.

GRUNDIG SATELLIT 700

At the time of going to press in October 1992, Grundig is selling off its Satellit 500 receiver at a discount price; just under 400 dollars depending on the dealer. This is in preparation of the launch of the new receiver, the Satellit 700. In Europe the 500 has already been replaced by production versions of the Satellit 700. Some pre-production models were reviewed at the end of 1991 with mixed results. There were some complaints that the first sets went into oscillation when the volume control was turned up full. Grundig said they were aware of these problems, fixed them, and then gave the OK to the factory in Portugal to start making sets for the shops.

At first glance the Satellit 700 looks similar to its predecessor, the 500. The styling of the case is the same, but the electronics inside are completely new. The case measures 30 by 17 by 7 cm and weighs 1.8 kilos without the 4 Size-D batteries which fit in the back. We think that puts the radio into the transportable category - not for the traveler who likes to keep things light and compact. If you want a radio that will deliver a bit of volume while you're on holiday, let's say in a caravan, this is one of the radios you should consider.

COVERAGE

Like most portables, you can use this radio for medium, long, and shortwave reception, plus FM stereo. The set has only one speaker, so if you want stereo you have to use headphones or connect an additional matching speaker which is available as an optional extra. longwave coverage is from 150-353 kHz on most versions, then there's a gap which is normally used by beacon stations. mediumwave coverage starts at 528 kHz and goes to 1611 kHz where shortwave coverage takes over, and that's continuous until 30 MHz, slightly lower if you buy the set in Italy where there are some legal restrictions. FM is between 87.5 and 108 MHz, and the set claims to be the first portable with the Radio Data System, which means that the set displays the name of the FM station it is tuned to, at least in certain parts of Europe.

TUNING

It is easy to switch on the radio and do some simple tuning. But if you really want to discover the secrets behind the on-board computer you'll have to sit down for a couple of hours and read the instructions. A glance at the instruction book shows where the emphasis lies. There are 32 pages in total. Half a page tells you the frequency bands for broadcast listening, a half page on a few specifications, one and half pages are used to explain the connection of an external antenna and single-sideband, half a page on fine tuning. The remaining 29 pages are needed to explain how to program the radio to do all kinds of scanning, copy memories, timer functions and so on. The German and Dutch versions of the instructions are much better than the English one, which has clearly been translated by someone who hasn't a clue about shortwave terminology. That's a problem because you need a clear set of instructions to get to know this radio. It took me about three days to get the hang of it all..... That's not a criticism as such....if you want these programing features, they're all there. The Satellit 700 is clearly aimed at the serious shortwave hobbyist. At press-time it was selling at around US$580 in Europe, whereas US dealers quoted us a price of US$550 (as from October 1992).

Most of the development costs in the Satellit 700 have apparently gone into the design of the software. But let's look at performance of the radio itself as an instrument for picking up distant radio stations.

EXTERNAL ANTENNA

If you use this radio at home, chances are you'd try to connect some sort of external antenna to the radio, even if its a piece of long wire dangling out of the window. The first thing you notice is that the internal ferrite rod antenna is always used for medium and longwave reception. You can't disable it, so you can't get rid of any in-house interference problems on medium and longwave, nor can you connect a directional loop. That we found surprising. When you switch in the external antenna, this becomes active above 1612 kHz.

We've often commented that receiver manufacturers tend to make portable radios that are so sensitive once you attach an external antenna they can't handle the incoming energy and simply overload. As a result you

get all sorts of whistles and stations popping up on parts of the dial where they are not really broadcasting. Measurements show that Grundig have solved this problem when you connect an external antenna to the 700. They've simply made the radio very insensitive! At 1.6 MHz you need 14 micro volts from the antenna to get a readable signal, at 4 MHz the figure is 3.7 microvolts, at 8 MHz sensitivity improves - 1.9 microvolts are needed, at 10 MHz 5 microvolts, at 20 MHz 2.5 microvolts. These are figures for the AM wide position 20 dB signal to noise ratio with 60% modulation. In the narrow AM filter, the sensitivity improves very slightly. Compare these numbers like 14, 4 and 2 microvolts with a communications receiver where figures of one microvolt are usually the norm, and look at the wide variation in sensitivity the Satellit 700 exhibits across the shortwave dial.

UNEVEN SENSITIVITY

The reason for this anomaly in the sensitivity lies in a preselector circuit which has to be peaked for maximum sensitivity. The radio does this peaking for you as you tune around, but clearly it doesn't track with much accuracy. You can peak the preselector manually using the keyboard, which sometimes improves the sensitivity quite a bit, especially around 15 MHz. But having peaked the preselector manually, if you then turn the main tuning knob slightly, the radio switches back to the default preselector position, and the extra sensitivity is lost. We couldn't find any way to tune across the bands with maximum sensitivity. Single-sideband results weren't that impressive either.

Sensitivity with the internal telescopic antenna is OK, better than one microvolt. That's because the signals picked up off the telescopic whip are put through a simple antenna preamplifier before being presented to the front end of the radio. Whilst the sensitivity goes up, the amplifier introduces all sorts of unwanted intermodulation products. Listening in Europe to the 20 meter ham band, around 14.2 MHz, you can hear all kinds of broadcast signals, and these are internally mixed signals from the 41 meter shortwave broadcast band. Not what you need for trying to dig weak stations out of the atmospheric noise!

The signal strength meter is simply a tuning guide. It doesn't correspond with real signal strengths because it doesn't start reacting until a very strong signal is heard.

AUDIO DISTORTION

Audio distortion is interesting to look at. In AM wide, the figure is an excellent 1.2%, in AM narrow the distortion is then 5%. When the synchronous detection mode is selected the distortion rises to 10%. On FM the distortion is very low. The internal speaker has a characteristic "German" sound when listening to shortwave. There's not much treble, but there's a lot of bass response. In fact the audio amplifier deliberately boosts the bass frequencies to get chest thumping results on FM stations. The bass tone control doesn't do much to remove this if you don't want it. On FM the Grundig Satellit 700 clearly outperforms both the Sony ICF-2001D and the Sony ICF-SW77.

SOFTWARE BUG!

The radio has an intelligent tuning software package, so you can tap in a frequency very quickly. There's one bug though. If you are listening to an SSB station on let's say 13630 kHz and you then type in 13640, the receiver assumes you want to switch back to the AM mode. You can avoid this by using the manual tuning knob instead. Fine tuning only works in the single sideband mode, allowing you to adjust the frequency plus or minus 150 Hz. The step in standard AM is 1 kHz on shortwave, and unless you switch to SSB or AM synchronous detection, you can't tune finer than that. That said, if you plan to connect a radio teletype decoder to the receiver, fine tuning in SSB won't be a problem. The automatic gain control is fair, but it seems to start working on signals that are too weak.

SELECTIVITY

Selectivity is the ability of the set to pick out the station you want, and reject the rest. The set gives you two bandwidth choices when listening to broadcast signals. In AM wide the bandwidth is 6.8 kHz at - 6dB, and in narrow 4 kHz at - 6dB. The shape factor of the filters is excellent. In practice the maximum selectivity is around 65 dB because the synthesizer in the receiver generates a fair bit of noise.

The synchronous detection system is an improvement over the Satellit 500, but, in our example, for some reason the detector is not symmetrical. The idea is that you tune in the station exactly. In Europe try tuning in Bayerischer Rundfunk on 6085 kHz. Then press the "sync" option on the keyboard. Now you can turn the tuning knob slightly to listen to either the upper or lower sideband, and in theory avoid strong interference from Radio Luxembourg on nearby 6090 kHz. When the bands are crowded that is often a method to escape interference from a station operating on an adjacent frequency. We could tune down as much as 1 kHz on the lower sideband side before the set lost its lock, but only 700 Hz in upper sideband, which is not enough to be really useful.

COMPARISON WITH LOWE HF-150

The synthesizer remains locked on even the weakest signals, but if the signals are really weak then noise and whistles from the synthesizer itself become audible. A comparison with the newly launched Lowe HF-150 show that the sync detector on the Lowe is superior. The Lowe receiver doesn't have a lot of FM features on it though, and none of the memory tricks of the Grundig Satellit 700.

The dynamic range of the Satellit 700 turns out to be only 66 dB. This is OK for a portable, but not good when compared to the performance offered by communications receivers designed for real long distance listening.

RADIO DATA SYSTEM FEATURE

We mentioned that the Satellit 700 has the Radio Data System (RDS) on it. But the full advantages of this new system of tuning haven't been utilised. As with all RDS sets you need an excellent stereo signal before the station name is displayed, so this isn't a new way to identify weak FM stations. You can ask the set to display alternative frequencies for the same network, but you have to store these manually into the memories. And it's a shame they didn't use the RDS clock information to give the radio a super accurate self correcting clock. Of course the RDS mode will only be of use if you live within range of transmitters that use the system..at present that means a select number of countries in Europe.

PRE-PROGRAMMED FEATURES

You can set a timer to record favorite programs while you're out if you have a suitable cassette recorder. The main frequencies of 15 major international broadcasters are already programmed in the radio, and you can put another 64 stations in yourself, each with up to 8 alternative frequencies. The radio remembers the mode, and bandwidth selection, but not the preselector setting. If that's not enough, you can lift a flap on the front of the set and install extra memory chips to increase the capacity to over 2048 stored frequencies. Some clubs in Germany are cooperating with Grundig and selling pre-programmed chips for a very reasonable price with the latest frequencies of broadcast and utility stations programmed into them.

POWER CONSUMPTION

The set consumes a lot of power. At comfortable listening volume the consumption is around 200 mA, so a set of 4 D cells will last 75 hours. You can also put in rechargeable cells and set the radio so that the cells charge up from the mains supply when the set is not in use. A fully charged set of nickel cadmium cells lasts about 20 hours, but it took the radio no less than 4 days before they were fully recharged! That's because the radio tricklecharges the batteries. Not very convenient if you happen to be on holiday and want to do a lot of listening.

CONCLUSIONS

It is difficult to summarize the Grundig Satellit 700 in a few words. It is a full feature radio with endless possibilities. Its performance as a radio though is only fair, and certainly for a few dollars more the Lowe HF-150 is a much better shortwave communications receiver. But the Lowe doesn't have FM, scanning functions, or RDS. It is too complicated for the casual listener, and the performance will disappoint some people who expect a portable communications receiver. But if you're looking for features, there's nothing on the market at the same price of around US$550. It is up to you to decide whether these features are worth the price. Side by side against the ICF-SW77 from Sony, we conclude that the Grundig gives much better performance on FM, but the Sony gives better results on shortwave above 5 MHz. More information from Grundig, Kurgartenstrasse 37, Fuerth/Bayern D8510, Germany. Tel: +49 911 7030. In the USA, Grundig has a distributor. This is Lextronix Inc, 3520 Haven Ave. Unit L, Redwood City CA 94063, USA. Tel: +1 415 361 1611 or in the USA only call (800) 877 2228. In Canada call (800) 637 1648.

ICOM IC-R9000

This radio has a price tag which varies a lot depending on where you buy it. It is cheapest in the United States (approx US$4700), but considerably more expensive in Europe where the price is in the region of £4080. Despite the price tag, the R9000 has found a niche in the market, wedged between the top end of the semi-professional market and the "budget" end of the professional monitoring range of radios used by government installations and the military. There are also some executives who reach retirement age and get a lump sum from the insurance company. These people want something to last for the rest of their life and they want the best. So Icom is not exaggerating when it claims that they can't make the R9000 fast enough.

For most of us, the R9000 will remain out of reach. Nevertheless, judging from reader response, there is a strong interest in how well it performs. At first glance, an ICOM ham radio transceiver called the IC-781 looks very similar to the R9000. The IC-781 has been around much longer, but our tests show the R9000 is not just the 781 without the transmitter section.

COVERAGE

Most versions of the Icom R9000 cover from 100 kHz, continuously through long, medium, and shortwave, VHF and UHF frequencies until coverage stops at 1999.8 MHz. The French version does not include the FM broadcast band between 88 and 108 MHz to avoid a certain import duty. In Holland, though, the full coverage version is available, and that is what we tested.

The focal point of the receiver is an orange coloured cathode ray tube. When you first switch on it shows the frequency the radio is tuned to, within a resolution of 10 Hz, and the tuning step that has been selected on the main rotary tuning knob. You can jump up the dial in 100 kHz steps at a time, or select any one of 8 other tuning steps down to the very fine 10 Hz increments. Icom uses the screen to simplify access to the radio's 1000 memory channels, its video-recorder style timer, and its extensive scanning system. Menu information on the screen gives you options, and you select them using the 6 function keys just underneath.

Television sets and computer monitors are generally regarded as one of radio's ardent enemies. Just put a radio near a TV and all you will hear is a continuous buzz. Icom have therefore done a lot of screening to ensure the monitor inside this radio doesn't radiate radio frequency noise. Indeed you can put a portable on top of the radio and not suffer any interference. There is some unavoidable radiation from the front of the display though, and that is not removed when the screen is put into a standby mode.

KEYBOARD DESIGN

Whilst the menus on the screen are simple to use, the keyboard on the set is poorly designed. We asked several people to try it out, and most complained that because the keys are small and close together its easy to punch in a wrong number. Nobody uses a calculator by standing up vertically and trying to punch in numbers. Likewise, the R9000 would be easier to operate if a separate accessory was available so you could tap in the numbers on a horizontal keypad. The keyboard software is rather user unfriendly too. Instead of making a separate kHz and Megahertz button as other companies have done on cheap portables, Icom makes no such distinction. So on mediumwave for instance, you cannot just punch in 747 kHz. You have you put in 0.747 MHz. That takes some getting used to, especially if you want to jump from one part of the dial to another quickly.

PERFORMANCE

But the main reason for buying such a radio is the performance. On that score, the Icom R9000 excels. The background synthesizer noise is so low that you immediately notice how quiet some portions of the bands really are. The lower the background noise, the easier it is to pick out the weak signals. And with a dynamic range of 102 dB using the narrow SSB filter, this radio is not seriously desensitised by powerful radio signals operating near the frequency you are trying to listen to. We were somewhat surprised to find that the R9000 has no synchronous detection capability. Sensitivity throughout the coverage, not just on shortwave, is more than adequate. @text:Figures below are for 10 dB S+N/N (12 dB SINAD for both FM modes) given in microvolts (50 Ohms: 60% mod AM, 100% SSB, 1 kHz tone.)

FREQ/MODE	AM	SSB/CW	FM	WIDE FM
0.1-0.5	3.1	0.5	-	-
0.5-1.8	6.2	1.0	-	-
1.8-30	1.1	0.15	-	-
30-999	1.4	0.33	0.5	1.3
1000-1240	2.1	0.33	1.1	4.0
1240-1300	2.0	0.33	0.5	2.0
1300-1600	4.2	0.64	1.0	4.1
1600-1999	5.6	1.0	1.5	5.7

The four selectivity options switch in sharp or wide filters which offer good shape factors, enabling you to pick out signals you want.

Tuning is greatly enhanced by a spectrum analyzer on the front panel. If you select this then you get a visual representation of 25 kHz either side of the frequency you are tuning. Transmitters are shown as a sharp peak. In a crowded broadcast band like 49 meters you get something that looks a mountain range that is constantly moving up and down as propagation affects the incoming signals. The resolution turns out to be adequate to spot weak signals next to strong ones. You simply twist the dial to hear what the small blip on the screen is actually saying. On higher frequencies you can widen the "window", otherwise the blips on the screen are too far apart. If you are someone who likes to bandscan, this device is brilliant, even more so because it remains activated while the radio is in the scanning mode.

1000 memories might seem too much for a shortwave listener. But on higher frequencies where signals are more spread out, logging marker stations is very useful. The radio can be controlled by a home computer if an optional interface is connected. There is suitable software on the market for you to set up very complex scanning procedures if you want to. There's an option for a built in voice synthesizer which you can use to announce frequencies on one channel of a stereo tape recorder, while the audio is registered on the other. The voice announces in English or Japanese.

The radio has a video output, so by connecting a computer monitor you can start searching for television signals. If you have an external radio teletype or packet converter, the text can also be fed back into the R9000 so press agency copy appears on the screen, although the text might be rather small for some older people. Ventilation of the radio is important, the back panel gets rather warm within half an hour of continuous use. As it weighs 20 kilos, it is not a good idea to carry the ICOM-9000 very far.

CONCLUSIONS

To summarize, the ICOM R9000 represents truly state of the art technology at a price. If your interests are confined to below 30 MHz, it is probably difficult to justify. But if VHF and UHF are also desired, this radio is represents good value, as you would otherwise have to buy two types of communications receiver to get the same coverage as the 9000. It does offer noticeably better performance both above and below 30 MHz than sets costing one third the price, though it's not perfect. We found an internally generated spurious signal on 574 kHz which gave slight problems to mediumwave reception in the vicinity of that channel. There were weak internally generated signals on 11670, 11755, 18543, 27815, but not strong enough to cause reception problems. The R9000 is clear a sign of the sets to come in the late 90's. Maybe a shortwave only version of the R9000, with a correspondingly lower price tag might be something for the manufacturers to aim for.

ICOM IC-R100

The requirements for a good shortwave communications receiver are very different to those of a good scanner. The R100 covers 100 kHz to 1856 MHz without a break, but we have divided our comments into performance above and below 30 MHz. Icom has packed a receiver with 100 memories, and three modes (AM, FM(narrow) and FM (wide) into a box just 150 x 50 x 181 mm. The set comes with a suitable mounting bracket for use in a car or a boat. The radio's speaker is mounted on the bottom of the cabinet, so if you use it at home on a table then four rubber feet (supplied) have to be attached to allow sound out from underneath the case.

The R100 runs off a car battery or a 13.8 volt (3 amp) DC power supply, available as an optional extra. The low voltage power cord can be bolted to the case (preventing it from coming loose), but the color coding of the supplied cable could have been clearer.

The number of function keys on the R100 has been reduced to a minimum with the result that many keys have two or even three functions. This takes some getting used to, but the instruction manual has been clearly written in this regard, and the receiver's software is very intelligent. The radio has a multi-functional back-lit liquid crystal display (LCD) which includes information on the chosen mode, the size of the tuning step, whether or not the attenuator is activated, and, of course, the frequency.

The rear of the set has no less than 3 antenna inputs, but this is needed because, of course, no single antenna can cover 100 kHz - 1800 MHz. The lower frequency portion (100 kHz - 50 MHz) has a SO 239 connector, there's an N connector for 50-905 MHz and another N connector for 905 - 1800 MHz. Because of the band activity, the changeover at 905 MHz is not as handy in Europe as in Japan. But the set has provision to power an external relay to switch one antenna across either one of the N connectors. The handbook does not tell you that the potential coming out of the receiver is +10 volts between 100 kHz - 50 MHz, 0 volts between 50-905 MHz, and +10 volts again between 905 - 1800 MHz. The middle pin is positive.

VHF/UHF CHARACTERISTICS DIFFERENT

The required specifications for a good shortwave receiver are very different from a VHF/UHF receiver. For instance, sensitivity below 1 microvolt on a shortwave set is not really desirable. With the high powers used by international broadcasters, even simple antennas are capable of delivering signals between 10 and 100 millivolt to the receiver's front end circuitry. The radio's delicate tuning circuits have to be able to pick out the weak signals whilst rejecting this enormous unwanted energy. On VHF/UHF the sensitivity becomes far more important. The limiting factor is not the atmospheric noise, but the internally generated noise from the front end! But in the busy communications bands (e.g. around 118 MHz) it is also important that strong signals on nearby channels do not desensitize (or "block") the weaker signals you want to listen to.

PERFORMANCE BELOW 50 MHZ

The R100 is clearly designed for broadcast listening below 50 MHz. Only signals in the AM mode can be demodulated, so if you want to listen to radio amateurs using SSB or teletype signals using Frequency Shift Keying (FSK), then you can immediately reject the R100. Amateur Radio

Exchange in Ealing London (Tel: +44 81 997 4476) recently launched the IC-R100SSB which offers an SSB capability. A similar offer is being made by Electronic Equipment Bank (EEB) of Vienna, Virginia.

The AM tuning steps are selectable at either 1, 5, 8, 9, 10, 12.5, 20, or 25 kHz. For shortwave, the 1 kHz step was slightly too coarse....100 Hz would have been better. The R100 does have an Automatic Frequency Control which locks onto VHF/UHF signals and compensates for any slight frequency drift by the radio as a result of temperature changes in the listening room.

The R100 has 100 memories, plus a further 20 which are used to determine the scanning limits, and one channel which is used for a priority frequency. Although the memories store useful info like the mode of the signal, they are not arranged in "banks". Scanning 100 channels takes far too much time. Whilst it is true that empty channels are skipped, in practice 40-50 important channels are possible when listening in a city area. When you only want to listen to a limited number of channels this means you have to program a skip marker in all the other unwanted channels! This is very time consuming.

Other sets offer a method to "bank" frequencies (say all aeronautical channels) together in one group. The R100 offers no less than 9 different methods of scanning the memories, skipping or only selecting certain modes, as well as automatically scanning the dial and storing active frequencies in up to 20 channels of the memory. But this flexibility is marred by a practical problem. When the receiver scans the dial it stops as soon as it hears a signal. Scanning resumes when the signal goes off the air....quite common on VHF/UHF communications. However if the set stops scanning because it has hit an internally generated birdy, or a transmitter that is on continuously, the scan is suspended. Most scanners allow you to jump over this problem with the manual tuning knob or the UP/DOWN buttons. If you try this on the R100 you jump out of the scanning mode. If you then switch to the scanning mode again, the radio restarts at the pre-programmed start frequency. There are ways round the problem by programming unwanted frequencies into the memory for the radio to skip, or scanning in the pause mode, but both solutions are clumsy.

The R100 has several timing functions similar to a video-recorder, but no way to remotely control a tape-recorder. The only solution is a cassette deck which starts when an audio signal is applied to the input (often termed VOX control).

SENSITIVITY

The IC-R100 is sensitive enough on shortwave, figures are around 1.5 μV (10 dB s+n/n, measured with 1 kHz tone, 60% modulation, into 50 Ohms). Above 50 MHz, where the atmospheric noise is much lower, sensitivity becomes more important. The R100 is also more than adequate in this region too. The preamplifier on the set, however, is not much use at all.

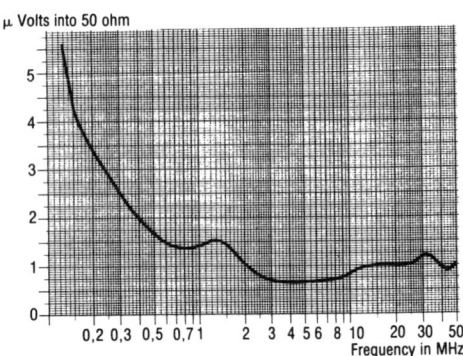

Icom R-100 Sensitivity: 100kHz–50MHz for 10db s+n/n AM 60% modulation depth measured, 1kHz tone.

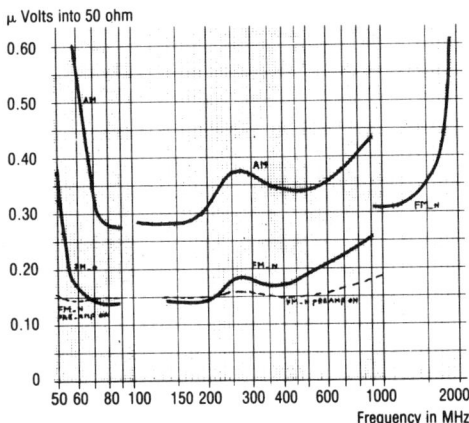

Icom R-100 Sensitivity: 50-1856MHz for 10dB s+n/n. FM narrow 4.8kHz deviation, modulation fequency 1kHz.

Ideally such a preamplifier should be in the antenna mast to be most effective. The preamplifier on the R100 only works in the 50 - 905 MHz region, whereas it would have been most effective above 905 MHz where the sensitivity of the front end drops off. The problem is that preamplifiers not only amplify the incoming signal, they add noise to the signal too. preamps designed for a small specific frequency range can achieve very low noise levels, but wide-band preamplifiers are much noisier. Between 50 and 144 MHz the preamp does not help in the reception of weaker signals because it generates the same noise as the front-end RF amplifier.

Between 144 -174 MHz, the sensitivity is even slightly worse with the preamp switched in. Above 220 MHz the preamp starts to show results, but because the dynamic range of the radio is quite low, overloading is quite possible. Leave the preamp switched off, and check whether the attenuator helps. Often switching in the 20 dB of attenuation results is a signal that's easier to understand!

SQUELCH

The squelch is useful on VHF/UHF to silence the set when no signal is being received. The receiver will only start to scan when the squelch is active, so it is important that it works well. If the R100 is mounted in the car, or there is fading on shortwave signals, the received signal can vary enormously. Some squelch designs have a so called "hysteresis" effect. Once the squelch opens for a reasonably strong signal, it will shut again at a much lower signal level, thus compensating for fading. The IC-R100 doesn't do this, so signals on the edge of the set squelch level cause an annoying "chuffing" effect. The upper signal limit of the squelch is fine for shortwave, and the AM/FM WIDE modes on VHF. But we found the lower limit too high.

SELECTIVITY

The manufacturer's specifications state that a 6 kHz filter is used for AM, and a 15 kHz filter for FM narrow reception. But this doesn't say much about the shape of the chosen filters. It is more interesting to look at how well strong signals in the same part of the band are rejected in favor of the signal you are trying to listen to. The so-called RF-protection ratio says a lot, and the R100 doesn't score too well. There are several reasons for this, not least the rather high synthesizer noise generated internally by the R100. This leads to reciprocal mixing, seriously affecting the dynamic selectivity. Also there may be some signal leakage around the ceramic filters. Whatever the cause, long wire antennas present so much signal to the input circuitry that the receiver can't cope. In a car the efficiency of the antenna is lower, but since the signals are weaker you are again faced with the basic fact that, below 50 MHz, the R100 is only really designed to pick up the stronger shortwave broadcast stations. The same story applies to VHF...the dynamic selectivity of this radio is well below what one would expect from a scanner in this price class.

The blocking level of the receiver on VHF is just 41 microvolts....signals stronger than this totally de-sensitize the receiver, even if they are up to 5 MHz away from the frequency being listened to. That will no doubt cause problems if you use the set in the vicinity of a FM broadcast station or a mobile telephone repeater.

The figures for 3rd order intermodulation products were also somewhat disappointing too. The radio has an intermodulation free dynamic range of just 69 dB. Signals that produce around 1.1 millivolts at the antenna terminals (easily done by broadcast signals) produce third order products which have a signal strength approaching 10 microvolts, which makes weaker signals unintelligible.

CONCLUSIONS

We could go further, but the bottom line is that this receiver is an example of a rather large compromise. In fact you're getting shortwave performance equivalent to SW receivers costing half the price. The same is true of the VHF/UHF section. If large frequency coverage is important, then this radio provides it. If you want to search for weak stations then you'd be better off deciding whether your needs lie either on shortwave or VHF/UHF and buy a shortwave receiver or VHF/UHF scanner to do the job. The radio seems to be available world-wide, with the exception of North America. The price in Europe is around £499. The scheduled price in the USA was US$679 before it was withdrawn from the market because of patent problems.

ICOM IC-R1

We could easily publish specification graphics for the R1 alongside those of the R100, for in many respects the receivers are similar. We decided against it because, frankly, the receiver is so poor some of the graphs would simply be meaningless.

SELECTIVITY AND SENSITIVITY

It is important to note that the sensitivity specifications have been measured at the antenna input of the receiver. The "rubber duck" antenna supplied with the radio clearly states it is for VHF/UHF. The antenna resonates at 130, 320, 425 MHz, so its efficiency below 50 MHz is very poor - nothing at all could be heard on the longwave broadcast band in Europe.

The receiver is sensitive enough. But the radio has a VERY noisy frequency synthesizer (contrast this with the IC-R72), so that all stations (even the strong ones) have a bed of white noise in the background.

The IC-R1 has one AM filter which, at 15 kHz, is much too wide for serious listening on medium and shortwave.

The graph shows that a station just 30 kHz away from the selected frequency cannot be more than 10 times stronger than the wanted station before all kinds of splatter and intermodulation products ruin reception. In our tests it was impossible to receive any signal without some sort of interference.

Replacing the inefficient "rubber duck" antenna with a longwire does nothing to improve reception. Indeed, signals at the radio's antenna input are so strong that the circuitry just blocks. We found evening reception of any station in the 49 or 41 meter broadcast band to be just a babble of mush. Assuming that this problem was the result of a bad sample, we purchased a second R1. The second set gave identical results.

Performance above 50 MHz is slightly better. The set is extremely sensitive...not surprising because the small antenna is not efficient. You don't need much of a signal to reach an acceptable signal to noise ratio.

But the selectivity is again disappointing. The filter for FM narrow is 15 kHz wide. The synthesizer noise once again plays a role so that stations between 30 and 100 kHz away from the wanted signal are only suppressed

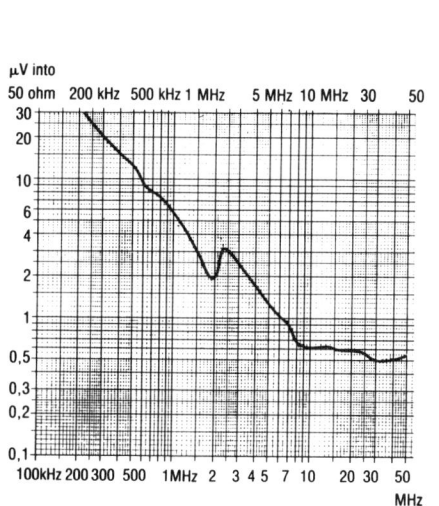
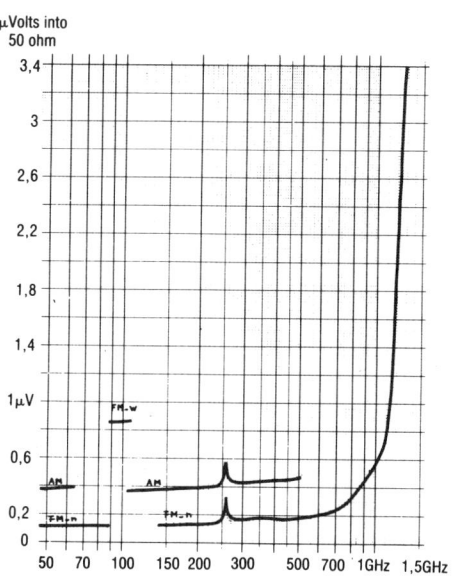

FIGURE 12: HF SENSITIVITY GRAPH   *FIGURE 13: VHF/UHF SENSITIVITY*

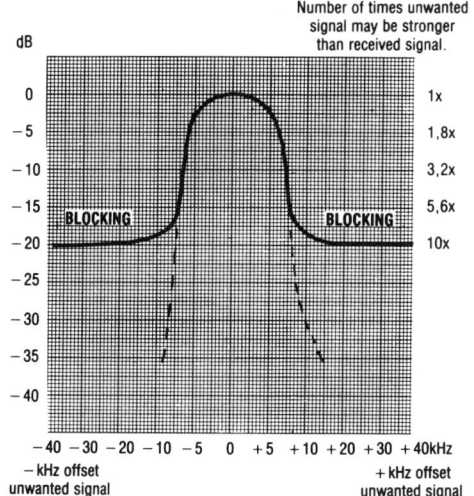
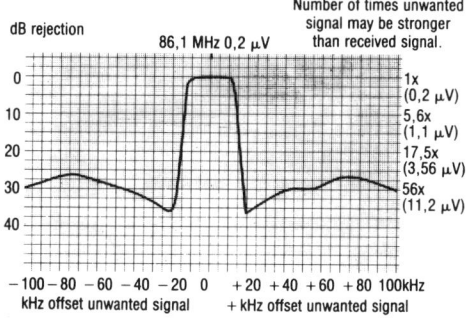

FIGURE 14: SELECTIVITY GRAPH R1   *FIGURE 15: VHF/FM SELECTIVITY R1*

by around 35 dB. Strong nearby stations (e.g. FM broadcast) cause severe intermodulation products that mask weaker signals in the police band. Only the stronger VHF/UHF signals can be successfully copied.

As with the R100, the R1 can only scan the entire 100 memory channels. They cannot be grouped into banks.

To conclude, it would seem this radio is a marvel of miniaturization. But the performance cannot match the looks. For the same money you can buy an excellent scanner, and have change to buy a shortwave portable for stronger international broadcasters. The R1 is a brave attempt, but it needs serious reworking in the lab. The radio seems to be available world-wide, with the exception of North America. In Europe the price is £369.

ICOM IC-R72

Communication receivers seem to have a sales lifetime of around 4 to 7 years, so just on cue ICOM in Japan launched the R72 to replace the existing ICR71. In fact when the 71 appeared it was an upgrade to the revolutionary ICR70 which was launched in 1982. The original R70 was a breakthrough in performance, but it was extremely difficult to operate, and there were problems with quality control.

The R71 has been a useful reference to measure other semi-professional communications receivers. It has excellent quality filters, and a very good dynamic range. Apart from some later examples, the R71 boasted a variable bandwidth control. This allowed the user to reduce the outer limit of the bandwidth, eliminating just enough unwanted interference without degrading the audio quality too much. However, on the subject of audio quality, the R71 has been infamous for its substandard audio quality, with distortion levels reaching 20% when used in the AM mode. Whilst this may be acceptable for ham radio use, it can be annoying for program listening.

The new IC-R72 will eventually replace the R71, but don't let the type number mislead. The new receiver has less features than the R71...but then it is also cheaper too! We tested an off-the-shelf model purchased in Europe during November 1990. We rechecked stock in August 1992 and found similar test results. Performance wise we found this new R72 has a lot going for it. The front panel measures 24 cm by 9 cm, whilst the unit itself is 23 cm deep. On the right of the front panel is the familiar key pad for direct frequency access, manual tuning knob, whilst less used controls appear as a row of tiny push buttons along the bottom of the front panel.

TUNING

The buttons of the keypad have a cushioned feel to them, the radio beeping at you to confirm that you've pushed that key. The tuning software appears to be similar to other Icom receivers, which isn't all that intelligent. Whereas you can punch in 198 kHz on many portable radios and the radio will move to longwave, followed by 25790 and the radio knows you mean a frequency in the 11 meter band, the R72 demands you either enter a decimal point to indicate MHz, or you are forced to enter a lot of trailing zeros. This takes some getting used to.

On the other hand Icom have improved the manual tuning by adding two buttons marked MHz and kHz. Normally the set tunes up and down the dial in 10 Hz steps. The clear back-lit liquid crystal display also has a 10 Hz resolution (this an improvement over the R71). If you then press the kHz button you start tuning in 1 kHz steps until you press the button again. If you choose MHz then turning the manual tuning knob changes the received frequency in 1 MHz steps. This is simple, and much better than so-called up and down buttons found on competing receivers. The receiver's tuning indicator shows when a station is accurately tuned in - based on the frequency readout, not the RF signal strength.

SENSITIVITY

The radio's sensitivity has been carefully chosen...not too sensitive on medium-wave to prevent overload by local stations. Then from 1.7 - 30 MHz the sensitivity is quite constant at around 1.2 µ volts, with the preamplifier switched off. The Automatic Gain Control characteristics are also carefully selected.

The LCD display shows the carrier frequency in AM, USB and LSB to within an accuracy of 70 Hz. In CW mode, howev-

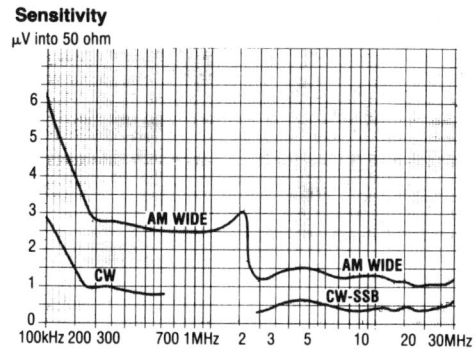

FIGURE 16: SENSITIVITY GRAPH

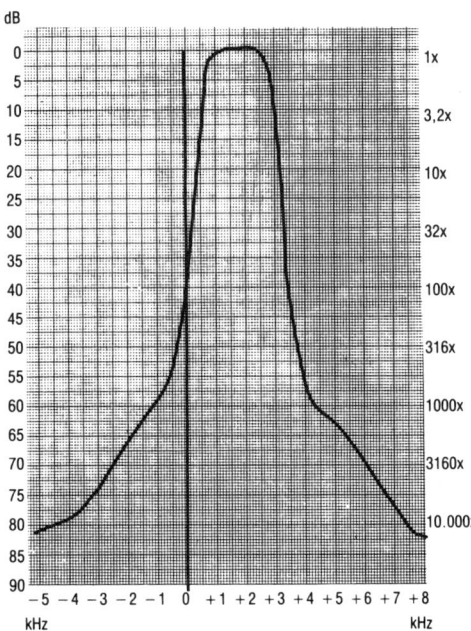

FIGURE 17: GRAPH OF SSB SELECTIVITY

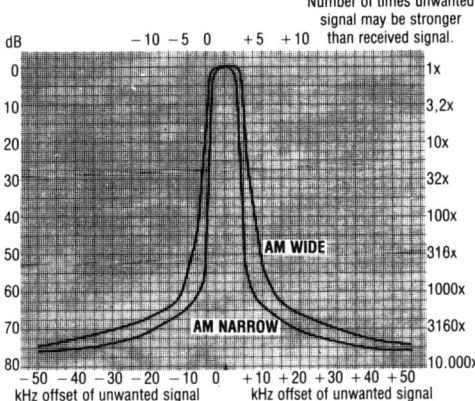

FIGURE 18: DYNAMIC SELECTIVITY GRAPH

er, the first sample tested displayed the frequencies 140 Hz too high.

The R72 receiver is different from the 71 model in that the voice bandwidth filters are no longer separate from the mode switch. So when you select upper or lower sideband, you have one bandwidth option. On the AM there are two, in the wide position the bandwidth is around 7 kHz at - 6 dB, 2.3 kHz in the narrow mode and on SSB. The AM wide filter is slightly too wide, but as the graph shows, both filters have very sharp cut off slopes.

The radio has a CW mode for the reception of Morse code using the 2.3 kHz filter. But for serious work it would be a good idea to install the optional 500 Hz filter.

GOOD DYNAMIC RANGE

The receiver has a reasonable dynamic range for its price. Using the AM wide bandwidth filter, and measurements at 9.1 MHz, the dynamic range free from intermodulation produced was measured at 83 dB. This is better than the Yaesu FRG-8800 and the Kenwood R2000, although below the (more expensive) Kenwood R5000.

On the back of the receiver are switches connected with the scanning functions (not so handy to put them there) as well as a high and low impedance antenna input. However, there is no switch between the two inputs. The manual warns you, correctly, not to try and connect two antennas or the overall performance will be severely degraded.

The radio has 99 memories, and it is easy to swap between a memory channel and the normal tuning mode. All this adds versatility. There is an optional interface which allows several functions to be remotely controlled by a home computer. The radio has a built-in timer than can be set to control a tape-recorder, although the internal relay will only switch a low voltage, not the mains current.

Back in 1982, the R70 was hailed by many reviewers for re-introducing the notch filter. In a perfect world, radio stations would all be exactly 5 kHz apart, but in practice some international broadcasters operate slightly off channel. This can lead to an annoying continuous tone. A notch filter allows you to remove that whistle, whilst letting the remainder of the audio frequencies through unscathed. The R71 continued the tradition of the notch filter, although it didn't work well in the AM mode. The R72's notch filter doesn't work at all because Icom have decided not to include one! A tone control on the R72 would have been useful too.

The radio does have a squelch control which can be set to silence the radio if a signal falls below a certain signal strength. The R72 has a scanning mode which allows you to automatically search a chosen portion of the dial. The radio stops scanning when it hits a signal that's above the level set on the squelch control. This feature is not new to communication receivers, although it's usually of more use on higher frequencies when the gaps between listenable signals are much higher, and the FM mode is in use.

Whilst there is considerably less distortion on the R72 than the 71 (typically 1.1% in AM at normal listening levels), the audio is only communications quality from the small built-in speaker on top of the radio's cabinet. If you plan to listen to radio stations for their content then you will get better enjoyment by connecting a pair of headphones or an external speaker. Icom offers this as an optional extra, together with a circuit board to allow the radio to operate in the narrow band FM mode, a voice synthesizer unit which announces the received frequency in either Japanese or English. If you live in an area where thunder storms are frequent, then there is an option to protect the delicate front end circuitry from damage by nearby lightening strikes. There is not much you can do against a direct hit.....so always disconnect the antenna during thunder.

The receiver's "S" meter on the sample we tested was accurately calibrated above S9, and below that value the meter gave readings that were slightly too low. But the scale was perfectly logarithmic up to S9 + 30 dB.

The R72 offers a two position noise blanker which we found to be occasionally effective on the ham bands but not much use for general broadcast listening. It marginally reduces the effects of ignition pulse noises, but at the same time it compromises the receiver's dynamic range. We found that by switching it to the high position and tuning to the 41 meter band here in Europe after the hours of darkness, all sorts of strange mixing products were noted...but these vanished when the blanker was switched off.

CONCLUSIONS

In short, the Icom R72 is neat compact communications receiver. It doesn't have a notch filter or a pass-band tuning control so it is less of a serious

DXer's receiver than the R71. But as an entry level receiver for the serious shortwave listener, this radio scores well. The radio seems to be available world-wide. The price for the IC-R72 in the UK is £589, a full £286 cheaper than the IC-R71! In the US the IC-R72 was launched much later, currently retailing at around US$830. In Japan (Akihabara district of Tokyo) we found the price to be 98,000 yen. But we have seen it priced in other areas much higher, so that the difference between it and the Kenwood R5000 (a really good "DX" machine) is quite narrow. Bear this in mind when making a decision.

JRC NRD-535

The NRD-535 is the latest offering from the Japan Radio Company. The styling of the new receiver is a clear improvement over the older 525, and JRC have worked hard to make the set easy to use. There are a lot of controls on the front panel, but you also get a lot of information in the display to tell you what's been activated. It's important to read the instructions before use.....you must take off the set's cover to remove the foam strip that prevents damage to the plug-in boards during transit. If you don't there's a good chance the set will overheat! The internal construction of the antenna is excellent, making use of plug in cards for easy maintenance.

TUNING STEPS

The manual tuning knob tunes in either 1 Hz, 10 Hz, or 100 Hz steps, depending on how a button is set. The UP-DOWN buttons shift frequencies in 10, 100, or 1000 Hz steps. The digital display can show frequencies to a resolution of 10 Hz.

SENSITIVITY

The 535 is a very sensitive receiver, more so than the 525, but this in fact doesn't mean you'll be able to get any more stations. On HF, at least, once you connect a reasonable antenna, the atmospheric noise reaches such a level that sensitivity much below 1 microvolt means that you're just amplifying the noise as well as the signal. The receiver has a built-in one-step attenuator, but at 20 dB it's somewhat coarse. It would have been nice to have a finer step in between,. e.g. 10 dB. Once again, with the high powers being used in the broadcasting business, the sensitivity figures are just academic.

The NRD-535 is specified to receive signals down as far as 100 kHz, although in fact it will tune lower than that. Below around 30 kHz you'll simply find noise from the receiver's internal synthesizer, but at the test site in The Netherlands strong reception is possible of the time signal station at Mainflingen, near Frankfurt Germany. That's on 77.5 kHz.

The following sensitivity measurements were made for 10 dB s+n/n measured in microvolts at 50 Ω. AM with 60% modulation.

FREQ RANGE kHz	AM WIDE	AM INTER	SSB,CW,RTTY (INTER)
25-100	NOT USABLE	NOT USABLE	NOT USABLE
100-470	0.8	0.5	0.32
470-1800	0.42	0.38	0.22
1800-30000	0.52	0.40	0.23

On the 525 there was a noise in the long-wave region which turned out to be caused by interference from the set's display. There were some modifications available from the ESKA company in Sweden to reduce this problem. However, on the NRD-535 the problem of noise generated by the display has become about three times worse. We found annoying levels of interference in the frequency range between 30 and 250 kHz. In a later version of the 535 tested in late June 1992, this problem had been reduced but not solved.

The display shows the carrier frequency in AM, SSB and in CW it compensates automatically for an 850 Hz tone. You can adjust the pitch of this tone if you want to through default settings. This compensation also applies to radio teletype.

ATTENUATOR AND RF GAIN

The attenuator is a relay switched 50Ω attenuator which unfortunately only has one step (20.6 dB). Use of the attenuator does affect the receiver's intermodulation performance. In addition though, JRC have provided an RF gain control which is continuously variable. The RF gain control does make the set much quieter, but doesn't affect the set's intermodulation performance.

PASSBAND, AGC AND S METER

The two AGC positions are fine for general listening, but we found the attack time too slow for fast scanning with computer control. JRC say they are aware of this point and it can be corrected.

The S meter is strange. Under S5, i.e. 3 microvolts, the "S" meter shows nothing at all. By the time you get to S9, the meter just shows S6. In both our examples, weak signals are always shown weaker than they really are.

The notch filter does work in the intermediate frequency section, but we could only reduce heterodyne whistles by 25 dB, which was not as much as we expected. The notch characteristics are quite broad. The squelch control is fine for most work, although the threshold level on both our examples was set rather too high. Signals were already easy to understand before the squelch cut in. The NRD-535 is a stable receiver. After a 30 minute warm up, both sets drifted less than 2 Hz in a half hour period.

DYNAMIC SELECTIVITY

The wide filter in the first batch of sets was too wide. It has since been reduced to 5.5 kHz which makes it much better for general broadcast listening and the lock of the synchronous detection is much better on weaker signals. An option of a third filter between the 5.5 and 2.6 kHz filters would be handy for broadcast DU work. In fact JRC have a variable bandwidth option which does the trick, but it's not included in the standard price

DYNAMIC RANGE

JRC quote a dynamic range of 106 dB in their brochure but they also say that they measured it with a bandwidth of 300 Hz. To get that you'd have to buy a special filter, and you certainly wouldn't use it for broadcast listening. We measured the dynamic range using the 2.6 kHz SSB filter and got a result of 88 dB. That's an good figure, but note that the older NRD 525 had a dynamic range of 92 dB . The 300 Hz filter gave a result of 98 dB, so we couldn't repeat JRC's 106 dB. The third order intercept point in the SSB mode, using a separation of 30 kHz, came out at +2 dam. We measured some birdies on the receiver, the strongest being at 97, 8450 and 8480 kHz. A click is sometimes noticeable as the set turns over a new MHz, e.g. at 2, 3 and 20 MHz.

ECSS OPTION

The NRD-535 has an option for Exalted Carrier with Selectable Sideband. Previously this was only available on the NRD 525 as a modification by the ESKAB company in Sweden. The new NRD-535 works better than the ESKAB modification, locking in around 2 seconds and staying locked until signals are just 3 dB above the noise floor. ECSS helps to improve problems with selective fading on weak signals, but in our examples the audio distortion went up from 2.7 to 5 %.

Lowe Electronics in UK have made a slightly souped up version of the 535, adjusting the filter characteristics, making slight improvements to the locking capacity of ECSS option and putting in a more powerful audio amplifier. That costs in the region of an extra £100 for the option.

The set comes complete with a RS232 output for a computer. We've tested both MS-DOS and Macintosh software which allow you to control a lot of the receiver's functions via an external PC. The programming language of the control commands is very easy to learn, so you could write your own control programs without much difficulty. The NRD-535 is by far the most versatile receiver on the market when it comes to software control, and it's a standard feature.

SOFTWARE UPGRADES

Since its introduction JRC has made quite a number of software modifications to improve the receiver, most notably allowing the bandwidth control option to operate in both the narrow and wide filter modes (originally it only worked in the narrow mode). If you are writing your own software to control the NRD-535 remotely, then make sure you are at least using software revision "G" or you may find that problems arise. Sets sold after March 1992 have this version fitted. JRC's New York office can help with advice on software revisions, and offers a "new chips for old" policy for a modest fee. Contact Paul Lannuier, JRC New York, 430 Park Avenue, New York, NY 10022 USA. Tel: +1 212 355 1180. Fax: +1 212 319 5227.

CONCLUSIONS

The price of a standard NRD-535 in Europe is around US$2000. The price in the US is considerably cheaper, namely US$1600. There are deluxe versions of the receiver with options fitted as standard. In Japan (Akihabara district of Tokyo) we found the price to be 138,000 yen for the NRD-535D version with all the options fitted.

We conclude that this receiver offers excellent value for the serious shortwave listener. Most of the specifications are excellent, the only slight reservations being in the area of intermodulation and dynamic selectivity. JRC have fixed the selectivity problem on receivers sold in 1992. The company has listened to criticism of the 525, and incorporated a lot of extras as standard (e.g. a computer interface). Since the NRD-535 is the same price as the old NRD-525 you're getting a receiver with more options for the same price, a timely upgrade which still gives JRC the edge. We have found JRC's New York office to be particularly knowledgeable about the receiver.

KENWOOD R-2000

This is one of the oldest shortwave receivers still on the market. Launched in 1983 it seems amazing this set has maintained a constant position in the marketplace for so long. It is wedged firmly between the higher end portables and the mid-priced table top models. Its longevity has probably been due to its straightforward operation and reasonable audio quality.

At a price of around US$650 you may want to check the set against the more sophisticated portables (e.g. Sony ICF-SW77) or pay a bit more for a Kenwood R5000 (MUCH better performance) or the Drake R8. The main problem with the R-2000 when compared with today's competition is the rather poor dynamic range.

COARSE TUNING

The receiver tunes 150 - 30000 kHz continuously in 50 Hz steps. These steps are rather too coarse for RTTY reception and ECSS. The lack of any form of receiver incremental tuning is a drawback. Tuning is via a conventional tuning knob or UP/DOWN buttons (the buttons moving in 1 MHz steps). There is no keypad entry. 10 memories is rather limited for a table top set, and so the scan-memory function doesn't offer enough scope 10 years on. Two filters are installed at 6 and 2.7 kHz. The skirt selectivity on the narrow filter is fine for specialized listening, the limiting factor being the restricted dynamic range. The WIDE filter is fine for listening to strong broadcast signals but any side channel splatter immediately causes problems.

RF GAIN CONTROL

The Automatic Gain Control has two positions (fast and slow), but you cannot switch it off. This seems a strange move since the inclusion of an RF gain control could have been useful for pulling weak stations out of the background noise. In practice the RF gain control is useful to reduce the overloading problems. The synthesizer noise is much higher than sets like the IC-R72 or Yaesu FRG-8800.

We tested an R-2000 in August of 1992, mainly for nostalgic purposes. We did our measurements using the 2.7 kHz filter with test signals 20 kHz apart. This gave a dynamic range of 82 dB. This is well above the CEPT minimum of 60 dB, and the very old R1000 (64 dB). But it is down on newer sets. Today's conclusion is simple. There are now better sets on the market for a lower price. A replacement for the R-2000 can't be far off if Kenwood wants to get back in the race at this price level.

KENWOOD R-5000

This set was launched in late 1987. The R-5000 is a table-top full coverage communications receiver. The light gray metal cabinet measures 279 by 107 by 307 mm, and it weighs 5.6 kg. An option exists to run the set off a 12 V car battery i.e. in a mobile home, but it is primarily designed for use in one location. Some versions are single voltage only (e.g. in Japan), though many will accept 120, 220, or 240 volts from the AC household current supply. The set uses its own internal re-chargeable battery to keep the two independent 24 hour clocks running and retain the memory contents when the set is not connected to household current. The receiver circuitry runs cool. Even after several hours use, the back panel of the receiver only gets lukewarm. This ensures excellent frequency stability.

TUNING

Official coverage of the receiver is between 100 kHz and 30 MHz continuously. An optional VHF module (installed in our example) allows additional tuning between 108 and 174 MHz. National legislation in some countries (e.g. Italy and Saudi Arabia) requires that coverage is more restricted. Check details to be sure. The test sample in fact tuned down to 30 kHz. Between 30 and 70 kHz nothing could be heard except synthesizer noise. But on 77.5 kHz the time signal station in Mainflingen, Federal Republic of

Germany could be heard with good strength at the test site some 500 kilometers away.

The receiver has two tuning modes, either via a manual tuning knob, or using a numeric keypad. Turning the manual knob in a clockwise direction causes the set to move up in frequency, the step being set automatically depending on the mode that has been selected. Normally this is 5000 Hz in the FM mode, 1000 Hz steps in AM, and 10 Hz steps in USB/LSB/FSK/CW. If a STEP button is switched on then tuning on AM, LSB, USB, FSK, and CW is in 100 Hz steps, 2500 Hz on FM. The steps mean accurate tuning of all modes is possible, including critical radio-teletype (RTTY) signals. For shortwave program listening, the 100 Hz step position is preferable when tuning. This avoids the "chuffing" effect audible on the larger step. For moving fast up and down the bands, the UP and DOWN buttons allow the user to jump 1 MHz in either direction.

Keyboard frequency entry takes a bit of getting used to. The mode and antenna buttons double as the keypad, arranged in two rows of five keys. This layout is different from a standard calculator. Unfortunately, the microprocessor software is not as intelligent as on competing receivers. Whilst a frequency such as 15560 kHz can be entered immediately, 6155 kHz has to entered as "06155", and 747 kHz as "00747". Further accuracy to the nearest 10 Hz is possible, e.g. 747.15, but not usually needed. The receiver's stability is excellent, allowing easy use of the Exalted Carrier Selectable Sideband (ECSS) receiving technique. The receiver is intelligent in that no retuning is needed when switching between any combination of AM, LSB, and USB. This is in contrast to the competing ICOM IC-R71 which jumps 3 kHz when swapping between USB and LSB.

TWO VFO'S AND MEMORIES

Operating convenience is enhanced by the use of two variable frequency oscillators. You can set one to a high frequency (say 15 MHz), and the other to a lower channel (e.g. 6 MHz). If you're using VFO A around 6 MHz and want to check conditions on higher frequencies, selecting the other VFO will allow you to jump to 15 MHz immediately. A 100 channel memory is also offered, storing the mode, and antenna number with each stored frequency. Selectivity positions are selected automatically, i.e. always WIDE for AM stations.

Scanning of the entire contents of the memory is possible, but also selected regions too. Channel numbers ending in 8 and 9 are usually reserved for setting frequency limits. Putting the receiver into "MEMORY SCROLL" mode allows the operator to scan all frequencies between the limits previously entered.

SIGNAL STRENGTH METER

International standards state that "S9" on a set below 30 MHz should correspond to a signal of 50 microvolts, and each S point below that is 6 dB lower. In general, "S" meters on communications receivers of this class only provide an approximate indication of signal strength. The R-5000 meter on our production model was fairly well calibrated. At 1.6 microvolts (i.e. S 4), the meter shows S 1 on both AM and SSB modes. Under S 7, the meter readings are too low, but at higher signal levels the meter is more accurate. An RF gain control plus an RF attenuator (10, 20, and 30 dB selectable steps) are offered for very strong nearby signals. We measured this and discovered that it provided exactly the amount of attenuation specified. Many receivers' attenuators do not. When you switch to VHF, however, the built-in converter provides extra amplification. As a result, the "S" meter becomes far more sensitive. The table below shows the results.

OFFICIAL VALUE (MICROVOLTS)	SHORTWAVE RANGE	VHF RANGE
S1 (0.2)	-	-
S2 (0.4)	-	S1
S3 (0.8)	-	S3
S4 (1.6)	S1	S3
S5 (3.2)	S4	S5
S6 (6.3)	S6	S7.5
S7 (12.5)	S8	S9 +2
S8 (25)	S9 +5	S9 + 8
S9 (50)	S9 +10	S9 + 15
S9 + 10 (158)	S9 + 22	OFF SCALE

FEATURES FOR THE VISUALLY IMPAIRED

Whilst not exclusively designed for the visually impaired, two features are available (one as an option) to help users who cannot see the receiver. If the VOICE button is pushed, a female voice announces the figures displayed on the blue fluorescent digital frequency readout. It is clear, but the output only comes through at a constant level in the headphones or the internal speaker. It is not available at the RECORD output. This is a pity. Using a stereo tape recorder, one channel connected to the receiver output, and one to the clock output, automatic logging would have been possible.

In addition, pushing the MODE buttons causes a single letter Morse code message to be heard in the speaker. (i.e. A for AM, L for LSB etc.). Clever thinking.

GOOD SENSITIVITY

In these days of 500 kW shortwave transmitters, sensitivity has become less important than good dynamic range. Nevertheless, checking to see how even the sensitivity of the receiver is over the entire range is interesting. The following results were obtained using the standard 10 dB signal to noise ratio.

All values gives in microvolts at 50 Ohms. FM mode measured at 1 kHz with 4.8 kHz deviation.

Frequency (MHz)	AM 6 kHz (60% modulation)	SSB Mode (2.4 kHz)	FM Mode (12 kHz)
0.07 -0.5	3.5	1	-
0.5 - 1.8	3.3	0.62	-
1.8 - 30	0.45	0.17	-
108 - 123	0.68	0.24	0.24
123 - 174	0.24	0.10	0.09

These are well within the manufacturer's specifications. Note, however, that sensitivity on mediumwave (also called the "AM Broadcast" band in North America) is much less than on shortwave. Kenwood have done this deliberately to avoid local mediumwave stations overloading the front end of the set. The sensitivity might cause problems if you wanted to use a mediumwave loop antenna for broadcast band DXing between 0.5 and 1.6 MHz. But in making a compromise, we feel that Kenwood has chosen the better solution. In the VHF area the R-5000 offers much better sensitivity that most scanners on the market.

We plotted the signal/noise ratio against the incoming signal. It is then possible to determine what level of signal is needed before an acceptable signal+noise/noise ratio is obtained. On shortwave, using the 6 kHz filter, the level was 3.45 microvolts for 20 dB S+N/N. Corresponding figures for the 2.4 kHz SSB filter were 0.9 microvolts.

GOOD SELECTIVITY

The R-5000 can be set to select the appropriate filter automatically, or a manual override is possible. The set comes with two filters NARROW and WIDE as standard for use in all modes except FM. The FM filter is 12 kHz wide at -6 dB. Because the medium and shortwave bands are very overcrowded, especially in Europe, the ability to separate the station you want from the rest is very important. The AM WIDE filter is within manufacturers specifications, i.e. 6.3 kHz at -6 dB, 19 kHz at - 60 dB. This is fine for general listening under minimum interference conditions. When the going gets tough, such as when DXing, it's likely that the operator will switch to the narrower filter. The results here are 2.7 kHz (-6 dB), 4.4 kHz (-60 dB).

Kenwood (and some other smaller companies) offer other filter options. The static selectivity of the 8.83 MHz crystal filter is good.

Various methods can be used to measure the selectivity of a receiver. Manufacturers, including Kenwood, usually quote results based on the curve of the intermediate frequency filter they've installed. This is a useful guide, but it doesn't guarantee how the receiver will perform in practice. European PTT administrations use a standard method to measure Maritime Mobile receivers for professional use. We applied the same test to the

R-5000. A signal generator connected to the receiver is set to 11500 kHz, and the output level set to cause a 20 dB signal to noise ratio. This is in the order of 1 microvolt on the R-5000. A second generator is set up with a 400 Hz tone and 30% modulation. Its frequency and strength is varied so that the 20 dB level on the receiver falls back to 14 dB. The difference in signal strength, measured in dBs (and often called the RF protection ratio), shows how much stronger a nearby transmitter needs to be before it causes interference. Likewise this is also a measure of how well a nearby signal is suppressed by the receiver's bandwidth filter. Optional filters exist enabling the user to select 270 Hz, 500 Hz, and 1800 Hz positions. These will appeal more to the Morse code (CW) and RTTY enthusiast.

DYNAMIC SELECTIVITY IN SSB MODE

When unwanted transmissions appear on -1 and +4 kHz (i.e. 1 kHz above and below the bandpass), the damping is 32 dB, mainly due to synthesizer noise. The CEPT norm is 40 dB. When the unwanted signals are at -5 and +8 kHz, damping is better at 60 dB (CEPT norm = 50 dB). The maximum suppression of unwanted transmissions operating on frequencies further away is 73 dB.

GOOD DYNAMIC RANGE

When two strong signals appear at the antenna input, mixing products occur in most receivers. They appear as "ghost" signals on other parts of the dial. These can be annoying, especially if a mixing product blocks out a weak signal you're trying to listen to. How strong these signals have to be before they cause problems is an important specification. The new 10 MHz amateur radio band is a source of weak signals near the strong international broadcasters around 9.9 MHz. Again using the standard CEPT measurements, the two signal generators were set to produce 0.961 microvolts, representing a signal 20 dB above the receiver's noise floor. The generators are tuned to 10000 kHz and 10020 kHz respectively, i.e. 20 kHz apart. Mixing products will occur at the same distance from both signals as their distance apart, i.e. at 10040 and 9980 kHz. The R-5000 is now set to SSB (using the 2.4 kHz filter), and tuned to one of the mixing products. The output of the two signal generators is now adjusted so that the level of the mixing product is 1 microvolt, i.e. a signal to noise ratio of 20 dB. Such a mixing product is not masked by the receiver's noise, and therefore is noticeable and annoying. The relationship between the initial and final levels of the signal generators is a measurement of the receiver's dynamic range.

In brochures, Kenwood quote measurements made at 50 kHz spacing and using a 500 Hz crystal filter. This gives 102 dB. We didn't install this filter, but by measuring with the 2.4 kHz filter and extrapolating, the figure does indeed turn out to be around 100 dB. This looks good on paper, but it doesn't help the customer compare the set with other receivers. We used the 20 kHz standard instead. This gave a 3rd order intercept point of

+16dBm, and a dynamic range of 90 dB. This is MUCH better than on the first pre-production R-5000 we tested. This is well above the CEPT minimum of 60 dB, and the old R1000 (64 dB). The ICOM IC-R71 comes out at 97 dB using the same method, the Yaesu FRG-8800 at 90 dB.

GOOD BLOCKING FIGURES FOR SHORTWAVE

Dynamic range is a useful measure of how nearby strong signals interfere with weak signals. "Blocking" is another test to see how nearby strong signals affect a strong signal being listened to. One standard test is to feed in a signal on 7100 kHz (arbitrary) of around 1 microvolt into the receiver, enough to give exactly 20 dB signal to noise ratio. The set is tuned to 7100 kHz. On 7300 kHz a second modulated signal is also applied to the antenna terminals. The level of this signal is turned up until a level is reached when all kinds of spurious background noises are heard on the R-5000. The level in this case was 35 millivolts (80 dB). This is a good figure and more impressive than 20 millivolts for the ICOM IC-R71 and 26 millivolts for the Yaesu FRG-8800. When the frequency spacing is changed to 1 MHz, the level rises to 44 millivolts.

TEN BIRDIES

Modern frequency synthesizers inside communications receivers are not perfect. Birdies are interference products which appear on the dial as silent carriers on AM, and whistles in the SSB/CW mode (hence the name). Weak signals on the same frequency as a birdie are subject to interference. They are measured by putting a 50 Ohm dummy load across the antenna terminals, putting the receiver in a screened Faraday cage, and tuning across the entire spectrum. Most birdies on the R-5000 are so weak that they disappear below the noise level when an antenna is connected. Ten birdies on the R-5000 are however above 1 microvolt. They are heard on 1977, 8546, 8999, 14053, 17999, 18069, 23076, 25909, 26999, 29865 kHz, but fortunately most fall outside the broadcast and amateur radio bands. This may seem like a lot, but in practice the R-5000 turns out to be a quiet receiver.

INTERMEDIATE FREQUENCY

Like most receivers in this class, the R-5000 is a double superheterodyne design. The first intermediate frequency for the R-5000 has been chosen as 58112.5 kHz, the second is 8830 kHz. IF Rejection of these two frequencies is better than 84 and 65 dB respectively which is good. The image ratio was measured as better than 90 dB, also good. Only when FM is used does a third stage at 455 kHz become active. This design is common to other Kenwood products.

MORE TUNING FEATURES

The receiver's automatic gain control can be switched between FAST and SLOW but not switched off entirely. An adjustable audio notch filter is also included in the set. It is capable of attenuating a single annoying heterodyne by some 31 dB, making it virtually inaudible. The notch is 270 Hz wide at -6 dB points. A useful pass-band tuning control is also offered, functioning in the SSB/FSK/CW modes only. The squelch control functions on all modes, but is primarily of use when listening to narrow band FM. Two independent noise blankers are included on the receiver. One is designed to eliminate the 10 Hz pulse caused by the former Soviet Over-the-Horizon Radar system, popularly known as the "Woodpecker". It is moderately effective in doing this, although the Woodpecker hasn't been heard on the air for a few years now. Other forms of pulse type noises can at least be reduced with the other blanker.

TAPE RECORDINGS

A built in timer can be used to control a tape-recorder for taping programs when the operator is absent. The internal relay can only switch low voltages though, not the AC household current supply. The RECORD level output is perfect for most tape-recorders.

COMPUTER CONTROL

An optional accessory exists, the IF-232C Interface, which fits between an RS232C interface on a home computer, and plugs into the back of the receiver. Kenwood tell us that software is up to the user and cannot be supplied by them.

CONCLUSIONS

The R-5000 is a good value communications receiver offering better performance than Yaesu FRG-8800 and comparing on equal terms with the ICOM IC-R71. The audio quality is excellent (much better than the ICOM IC-R71). The controls on the receiver take some getting used to. Some might judge them as rather clumsy. The receiver's moderate level of synthesizer noise is a slight drawback. But all the facilities you need on a communications receiver are available. In Europe, Kenwood have lowered the price of the R-5000 in some markets, putting it well below the ICOM IC-R71.

Average current prices = 2799 Guilders in The Netherlands, £895 in UK, DM 2050 in Germany and US$880 in the United States.

LOWE SRX-50

This receiver was launched by Lowe of the UK in July of 1992. The receiver is made for Lowe in the Peoples' Republic of China. That's why Lowe have given it an "SRX" type number to distinguish it from the "HF" series which they make themselves. Coverage is from 153-281 kHz (longwave), 531-1602 kHz (mediumwave) and 5900-15500 kHz shortwave. FM is offered between 87.5 - 108 MHz, stereo when using headphones. The set measures 18 x 11.5 x 3.5 cm, so pocket book size.

When switched off the display shows the time (24 hr format). Turning the radio on a red LED lights in the middle of a circle of 5 memory buttons. There is no keypad for direct frequency entry or a rotatable tuning knob. In theory you simply have to use the tear shaped up and down keys. In practice if you store three channels in the memory at the bottom, middle and top end of the tuning range you can jump quite quickly around the dial.

Audio quality for a portable turned out to be surprisingly good, and the single bandwidth filter is remarkably sharp. A light is included for reading the LCD display in the dark. Tuning on mediumwave is in 9 kHz steps and there appears to be no way to change this to 10 kHz when travelling in North America. Tuning on shortwave is in 5 kHz steps which is adequate for a radio of this type...there are only a few shortwave stations using split channels.

The biggest drawback with the receiver is the limited shortwave coverage. A glance at Chapter One shows that the new WARC bands cover a lot more than the range 5900-15500 kHz. Whilst the bit missing from the 49 meter band is not too serious, the fact that all channels above 15500 kHz are missing is a serious omission. A few more kHz and this receiver would qualify as the best portable on the market for £40. The synthesizer noise is certainly lower than both DAK receivers. Maybe the original equipment manufacturer can be persuaded to come up with the SRX-55. There is still a market for a low-cost digital portable that doesn't overload as soon as the sun sets.

LOWE HF-125

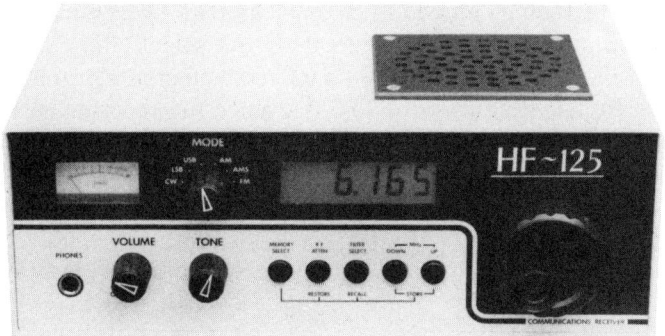

In the first part of 1987, a new shortwave communications receiver came onto the European market. Designed and built by Lowe Electronics of Derbyshire England, it is lightweight at 1.8 kg in its standard version. It measures 25 by 10 by 20 cm making it very compact. It is a double-superhet design, with a high first intermediate frequency of 45 MHz. The set comes with a 12 volt DC power adapter for use on 220 volts AC current in Europe. Although this is an external unit, the fact that the transformer is outside the radio doesn't lead to hum problems as experienced with other radios that have tried this in the past. Although now replaced by the HF-150 and HF-225, we have included it because the receiver is still seem in some retail outlets (although scarce).

The beginner will have no problems connecting up an antenna, a pair of headphones or a tape-recorder. Lowe Electronics have written a clear, logical instruction manual plus a booklet that explains what to listen out for when you start tuning around. Part of its attraction is the simplicity. There are just 9 controls on the front panel, which can be tilted up to face the operator. Up to 30 frequencies can be stored in the radio's memory.

TUNING

A large clear back-lit liquid crystal display shows the frequency being received to the near 1 kHz. Frequency display shows the true carrier frequency, whatever the mode selected. The single Phase-Lock-Loop VFO is switched three times throughout the range, at 6, 12 and 21 MHz. Reception on exactly these frequencies is blocked. The manual tuning knob offers FAST and SLOW tuning, depending on the speed with which the knob is rotated. The tuning increments vary with mode selected as follows:

Mode	Normal Speed	Fast Speed
LSB, USB, CW	15.6 Hz	250 Hz
AM	62 Hz	500 Hz
FM	125 Hz	500 Hz
AM sync	15.6 Hz	No fast mode available

The AM "normal" steps proved to be small enough for comfortable tuning without any annoying "chuffing" effects. If you select either upper/lower sideband or the CW mode as you'd do for radio amateurs or telex stations, the slowest tuning step drops to 15.6 Hz. Since the receiver is stable, the fine tuning steps allow you to accurately tune in a radio-teletype station with ease. The receiver also has a narrow-band FM option for monitoring citizen's band on 27 MHz for instance, but we didn't test this mode ourselves. The increment chosen for the AM sync mode is fine, but there is no provision for selecting either upper or lower sideband. This is a rather serious omission.

STABILITY AND BIRDIES

We put the receiver in a constant temperature environment (20 degrees Centigrade) and tuned it to 12100 kHz. The set was then switched to the USB and LSB modes. After an hour, drift was less than 50 Hz in each instance. This is comparable with receivers in much higher price classes.

If you put a shortwave receiver inside a metal cage so that it's screened off from the outside world, you can then measure "ghost" signals on the dial generated by the set itself. They appear as a silent carrier if you're listening on AM or as a whistle if you're listening on single-sideband. Most sets of this price range suffer from these so-called "birdies", and you can't get rid of all of them without drastically increasing the price of the radio. The designer's aim is to make them as weak as possible and on frequencies that are outside major broadcast and amateur bands. We measured more than 40 low level birdies on the HF-125 (i.e. greater than 0.5 microvolt). Most of these disappear below the noise level when you connect an antenna, but we found 10 "birdies" which were above 1 microvolt, and therefore noticeable. A few fall inside broadcast and amateur radio bands. We found ones on 143, 1500, 1945, 3141, 4901, 9891, 10793, 11954, 13935, 14082, 14482, 14505, 14847 (S5), 18749, 21587, 23998, 28784, 29695 kHz. If you try to tune in a station on these channels, you will get extra distortion or interference generated by the HF-125 itself. More expensive sets like the Yaesu FRG-8800 and ICOM IC-R71 have birdies too, but not quite as many. For a receiver in this price bracket, the number of birdies on the HF-125 is not abnormal, but tending towards the higher side.

SENSITIVITY

The sensitivity of a receiver is less important these days. shortwave stations using hundreds of kilowatts are able to deliver quite a strong signal into many target areas, especially Europe and parts of Asia.

In our tests we checked how much signal is needed at the 50 Ohm input to get a signal + noise/ noise ratio of 10 dB. The speech or music plus the receiver noise is then around 3 times louder than the receiver noise on its own, and just at the point of being intelligible. Sensitivity often varies with frequency, and for that reason the HF-125 was measured at several points in the spectrum. We checked the AM mode (using 60% modulation) and SSB. The 4 kHz filter was selected for the measurements, being the filter than SW listeners are most likely to use for broadcast listening.

Sensitivity in microvolts at 50 Ohms: 10 dB S+N/N, 60% mod AM, 100% SSB, 1 kHz tone.

FREQ (MHz)	0.2	2.2	8.0	12.1	16.0	26.0
AM	0.70	0.66	0.71	0.73	0.67	0.70
SSB	0.23	0.21	0.23	0.24	0.21	0.22

The receiver's built-in synthesizer generates a lot of noise between 30 and 70 kHz, way down in the very low frequency part of the spectrum. This masks most of the signals coming in. But from 100 kHz up to 30,000 kHz, covered continuously on this set with no breaks, the sensitivity is remarkably uniform. Sensitivity is also dependent on the chosen bandwidth filter. The following measurements were taken at 12.1 MHz.

Sensitivity in microvolts, measured at 50 Ohms:
10 dB S+N/N against bandwidth AM 60% at 12.1 MHz.

IF BANDWIDTH (kHz)	2.5	4.0	7.0	10.0
SENSITIVITY	0.52	0.73	0.75	0.80

The receiver offers an analogue signal strength meter which turns out to be very accurately calibrated, far more than just a tuning gimmick. Also of interest is the minimum antenna voltage needed to detect a signal above the receiver noise, and at what level you can easily understand it (i.e. 20 dB S+N/N). We used the 4 kHz filter.

Sensitivity AM 60%	12 MHz S+N/N3 dB :	0.22 microvolts
Sensitivity AM 60%	12 MHz S+N/N 20 dB :	2.45 microvolts
Sensitivity SSB 1	2 MHz S+N/N 3 dB :	0.10 microvolts
Sensitivity SSB	12 MHz S+N/N 20 dB :	0.89 microvolts

RF ATTENUATOR/ AUTOMATIC GAIN CONTROL

A push-button attenuator is provided, offering around 17 dB attenuation when measured at 12 MHz. It consists of a resistance-network, which can be shorted out of circuit by a diode. After the initial attenuation is seen on the S-meter, there is no further indication that the attenuator is switched in.

The AGC is designed to keep the audio level at a fairly constant level, even though the signal may be fading. Using AM 60%, 1 kHz tone, at 12 MHz the audio level varies by 3 dB for antenna input between 0.7 microvolts and 120 millivolts. The speed of the AGC is not selectable, and we found the time constant chosen to be rather slow.

STATIC SELECTIVITY

The HF-125 offers 4 different selectivity positions. There's a very wide 10 kHz setting, best suited when you're listening to local mediumwave stations, and then the low distortion audio (less than 3%) sounds very pleasant indeed to our European ears. The 7 kHz setting is best suited to shortwave when you've found a strong clear signal. The 4 kHz option is more useful when the signal you're listening to is suffering from strong stations operating on frequencies 5 kHz either side of the one you're trying to monitor. And finally, if you're trying to dig something out of the noise, or looking for utility stations, the 2.5 kHz filter is the best suited.

The first IF filter at 45 MHz has a fixed bandwidth of 15 kHz. In the second IF, use is made of a 14 element ceramic filter with a bandwidth of 2.5 kHz, and a 6:60 dB shape factor of 1:2. This is above average for the price range. The other filters in the second IF are low-cost ceramic types which are selected in various sequences to obtain the desired results.

DYNAMIC SELECTIVITY

The figures given above show the bandwidth of the filters themselves measured at -6 dB. But that doesn't give an indication as to how strong an unwanted signal has to be before it causes interference. Signal leakage near the filters can sometimes take place between unscreened components on the printed circuit board. Noise from nearby oscillators can lead to reciprocal mixing. As a result, the WRTH equipment lab tests how signals on nearby frequencies are rejected. This is called dynamic selectivity, or the RF protection ratio. The smaller 2.5 kHz filter clearly has much better dynamic selectivity than the 4 kHz filter, with poorer audio fidelity of course.

INTERMODULATION SUPPRESSION

When two strong signals appear at the antenna input, mixing products occur in most receivers, appearing as unwanted signals on other parts of the dial. These can be annoying, especially if a mixing product blocks out a weak signal you're trying to listen to. How strong these signals have to be before they cause problems is an important specification. The new 10 MHz amateur radio band is a source of weak signals near the strong international broadcasters around 9.9 MHz. The third order products are generally the most annoying. If transmitter A is on 10000 kHz, and transmitter B on

10020 kHz, then third order interference products may be found on 9980 kHz, i.e. 20 kHz below the 10000 transmitter, and 10040 kHz or 20 kHz higher than the 10020 kHz transmitter. In practice, the shortwave bands deliver far more of these examples and so it is important that these unwanted mixing products are not heard. Some receivers easily generate intermodulation products. In the case of the HF-125, when using the 2.5 kHz filter, both signals must each have a level of 15.5 millivolts before intermodulation products reach an annoying 1 microvolt. This is a very good figure.

DYNAMIC RANGE

Dynamic range is a measure of how large the difference in signal strength has to be between a weak and strong signal, before interference by the strong signal occurs. This is important if the receiver is to be used for DXing. Again using the standard CEPT measurements, the two signal generators are tuned to 10000 kHz and 10020 kHz respectively, i.e. 20 kHz apart. Mixing products will occur at the same distance from both signals as their distance apart, i.e. at 10040 and 9980 kHz. The HF-125 is now set to SSB (using the 2.5 kHz filter), and tuned to one of the mixing products. The level of both signals is now set so that the mixing product is 3 dB above the noise floor (0.1 microvolts on the HF-125). The output of the signal generator on 10000 kHz is now adjusted to the level of the mixing product. The relationship between the level of one of the two signal generators and the strength of the mixing product (in dBs) is the receiver's dynamic range. For the HF-125 this turns out to be 78 dB. This is well above the CEPT minimum of 60 dB, and the Kenwood R1000 (64 dB) but not as good as more expensive receivers. The ICOM IC-R71 comes out at 97 dB using the same method, the Yaesu FRG-8800 at 90 dB (adjustments made for different bandwidth filters). For a low-cost receiver though, the results can be classed as fair/good.

BLOCKING

The RF protection ratio is a useful measure of how nearby strong signals interfere with weak signals. "Blocking" is another test to see how nearby strong signals affect a strong signal being listened to. One standard European PTT test is to feed in a signal on 7100 kHz (arbitrary) of enough level to give exactly 20 dB signal to noise ratio. The set is tuned to 7100 kHz. On 7300 kHz a second modulated signal (400 Hz tone, 30% mod) is also applied to the antenna terminals.

The level of this signal is turned up until a level is reached when the signal to noise ratio of the desired station has dropped back to 14 dB (fair interference) or when the audio drops by 3 dB (whichever occurs first). The measurement is repeated with the second generator on 8100 and 6100 kHz.

Blocking: 20 dB S+N/N -> 14 dB S+N/N or -3 dB audio output
Desired Signal for 20 dB S/N

Unwanted Signal at 200 kHz	Unwanted Signal at 1 MHz
2.45 microvolts 66 dB (4.9 millivolts)	70 dB (7.7 millivolts)

These figures are quite fair for the price paid. A lot of whistles were noted on the HF-125 when signals between 20 and 200 kHz away were more than 60 dB stronger than the desired signal.

NO ACTIVE ANTENNAS

The Lowe instruction book warns against connecting huge antennas. We found an active antenna such as the Datong AD-270 or Dressler ARA 60 was totally unsuitable with the HF-125 in Europe.

PRICE

In its day the HF-125 cost £375 including Value Added Tax in Britain. For an extra £59.50 you can plug in a keyboard, and then enter frequencies directly into the set. The software here is friendly, simple and logical. Although the synchronous detection system does perform, it doesn't lock onto very weak signals, and when this mode is used, the audio sounds less pleasant. That's very subjective though. But unlike the cheaper SONY ICF-2001D, which also offers synchronous detection, you cannot select between upper and lower sideband. We would rate the value of the synchronous detector as only FAIR/POOR. The sync on the new HF-150 is MUCH better.

CONCLUSIONS

It's refreshing to find a receiver that does exactly what it claims. The manufacturer's specifications check out perfectly, and Lowe Electronics have achieved all the aims they've set themselves. This set has been replaced by the HF-150. The new receiver offers a much better synchronous detector and performance for a lower price. So bear this in mind when examining the second hand market for the HF-125.

LOWE HF-150

Holland and Britain seem to be the main marketplace for Lowe receivers so far. This English company started making radios to its own design and specification a couple of years back, with the launch of the HF-125 and HF-225 receivers. The HF-125 has since been discontinued. The approach has been to try and make a simple to operate radio with the best specifications for the price. Compared to Japanese competition the radios look quite plain, but performance wise they score well. At the end of 1991 Lowe announced it was launching a small receiver for the bottom end of the market as an alternative to Japanese push button portables....sets like the Sony ICF-2001D. In March 1992 we tested an off-the-shelf example of the HF-150, putting it through a series of laboratory and practical listening tests.

For a price of £329 in Britain (including VAT), you get a table-top communications receiver which at first glance looks surprisingly small. The case is made of metal, not plastic, and measures just 185 by 80 by 160 mm. It's quite light too, just 1300 grams without the 8 penlight batteries which fit into two special holders at the back of the set.

SIMPLE CONTROLS

From the front there are just 5 controls....a combined on-off switch and volume control, three buttons which have several functions including the selection of the mode and memories, and a large tuning knob. A large 5 digit liquid crystal display shows the frequency you're tuned to within the nearest kHz, If you push a button the display gives you information about the receiver mode and memory number, but normally it shows only the frequency, and there's no light to illuminate it. That's it. Lowe sell a keypad as an optional extra that plugs into the back of the set and you place in front of the radio as you use it. That's essential if you want to move quickly about the dial...other wise you have to move up and down in frequency by spinning the tuning knob. Getting from 30 kHz right up to 30 MHz, which represents the full coverage of the set, could take some time.

The set has no signal strength meter, you can't add extra filters at a later stage for very narrow bandwidth reception of Morse code, there's no notch filter, no noise blanker, and no tone control. But if these are extras that you can miss, then what Lowe have put inside the box turns out to be very acceptable indeed.

PERFORMANCE

The dual-conversion super heterodyne design is quite straight forward. Signals come in from an external antenna. You can switch in an antenna amplifier if you're using an indoor whip, although in practice we didn't need that at all in this part of Europe where signals are always strong. Signals go through a 30 MHz low pass filter before they hit the mixing stage of the radio. Of course there's a lot of energy coming off most shortwave antennas, bearing in mind the powers used by broadcast stations. More expensive radios use a series of filters to make sure that if you're listening to 15 MHz shortwave for instance, strong mediumwave signals, or stations in the 41 and 49 meter band are attenuated before they get to the mixing stage of the radio. Too much energy at the front end of the sensitive input circuitry can lead to overloading, and the appearance of signals on the dial which are the result of mixing products inside the radio. Having said that we measured the intercept point as +3 dBm using two signals 30 kHz apart. This gives you a dynamic range of 86 dB which is a fair-to-good value for a radio of this price. We disagree with the instruction book though, that recommends a long wire of up to 30 meters. Our tests in Holland showed that if you connect a wire longer than about 12 meters, you get enormous overloading problems once the sun sets. That will be less of a problem in low signal strength areas such as the Pacific or the American Mid-West.

ATTENUATION TRICKS

It's often assumed that the more signal you pump into a radio, the more distant stations you'll be able to hear. Well that's not the case. We found that late at night, weak and difficult signals were more intelligible if you switch in the 20 dB of attenuation. But that control is on the back of the set which is not easy to get at. If you use the set in Europe you might want to consider a separate aerial attenuator which say steps of 6, 12, 20 dB of attenuation, and give it a try on weak signals.

SENSITIVITY & MODES

We measured sensitivity using a signal modulated at 60% using a 1 kHz tone. We found that our measurements corresponded well with the results given by Lowe in their instruction book. Between 50 and 500 kHz the sensitivity is around 1.8 micro volts, and around 0.8 micro volts for the medium and shortwave part of the dial. There's not much difference in sensitivity between the wide and narrow filters used in the HF-150.

The radio has various modes. USB, LSB, standard AM, and you can also use what's termed synchronous AM. Unlike other Lowe sets available until now, the HF-150 allows you to listen to either the upper or lower sideband of a broadcast signal whilst in the "sync" mode. That's extremely useful when there's a strong interfering station 5 kHz away from the station you're trying to listen to. You can also use synchronous detection to reduce at least some of the effects of shortwave fading. The use of synchronous detection though in the double-sideband mode leads to some slight loss of sensitivity, but that's nothing to be concerned about. The background noise also rises slightly on the example we've tested. The radio takes up to two seconds to lock onto the desired signal, but once it's locked the radio does an excellent job of keeping in lock even when the signal fades to almost nothing.

HIGH BATTERY DRAIN

Battery consumption of the receiver is quite high, especially when compared to similar priced competition, anything up to 275 mA at full volume. We put in a set of 8 fresh alkaline batteries and got the radio to work for just 6 hours before they were flat. You can purchase rechargeable nickel cadmium batteries. When the set is switched off they automatically charge up. It takes about 16 hours to get a full charge after which you can use the radio for portable work for about 3 hours before you need to recharge again. The cheapest solution of all is simply to use the supplied external AC adapter which plugs into the back of the radio and gives all the power you need without any hum problems. Be careful if you use alkaline batteries and the mains adapter. Although there is a warning in small print in the instruction booklet, if you use the mains adapter while the alkalines are still installed, the HF-150 will try to charge these up when you switch off the receiver. Since alkalines are not designed for this they start to heat up and there is a danger that they might explode. We don't think that Lowe explain this clearly enough (there should be a sticker in the battery compartment!).

The HF-150 has two filters which have a bandwidth of 7 and 2.9 kHz respectively. These ceramic filters have a good shape factor for the price paid. So if the signal you want to listen to is strong you can really sit back and listen to the programming. The design of the automatic gain control is excellent, so no unwanted pumping of signals. The signal distortion is very low for a radio of this type, and if you connect the radio to a hi-fi set you'd be surprised what fidelity you can get out of a strong shortwave broadcaster.

PR-150

In mid-August 1992 we tested a prototype preselector designed to match the HF-150. This will be available in October 1992 at a price around £160. Initial results were disappointing. Although the unit reduced some of the second order intermodulation products it introduced its own third order problems. The PR-150 was handy when tuning the tropical bands, but offered no or even worse performance when listening to broadcast stations around 15 MHz. The set has a one step attenuator (-16 dB) right across 150-30000 kHz. The preamp offers an average boost of 7-9 dB depending on the section of the band being tuned.

If you can ensure that the front end of the radio only receives frequencies on or around the part of the dial of interest, then the problems of intermodulation can be reduced. For example, on simple receivers you often hear a lot of extra noise around 14 and 15 MHz which results from mixing products from the 7 MHz broadcast band. The preselector is fine for reducing this. But its own internal circuitry causes mixing products to appear from nearby channels. We measured the third order intercept point as -5dBm using signals at 8 and 10 MHz. If you're using a high quality active antenna such as a DX-7 from RF Systems (third order = 25 dBm) then putting the preselector between the receiver and such an antenna will actually make things worse when listening above around 12 MHz. The PR-150 has a definite advantage if you want to use the HF-150 for monitoring weather fax stations on VLF frequencies.

PR-150 BANDWIDTH

The preselector divides the bands into 9 distinct sections. Lowe are quite conservative in what these cover, thus ensuring that there are no gaps. Our measurements gave the following bands:

Bands in MHz of the PR-150

Band 1	13.3-31.7
Band 2	7.4-17.5
Band 3	4.26-10.2
Band 4	2.42-6.0
Band 5	1.44-3.25
Band 6	0.79-1.81
Band 7	0.43-0.98
Band 8	0.2-0.49
Band 9	0.09-0.2

With the exception of the first band we took the middle frequency of each range and measured the bandwidth at that point. Note how the bandwidth gets much wider as the frequency rises.

Bandwidth PR-150

Frequency (kHz)	-6dB point	-20 dB point	-40 dB point
22000	2.5 MHz	6 MHz	-
12500	2.0 MHz	5 MHz	-
7100	1.0 MHz	3 MHz	-
4100	400 kHz	900 kHz	> 2 MHz
2300	300 kHz	650 kHz	2 MHz
1300	125 kHz	300 kHz	1 MHz
700	40 kHz	120 kHz	380 kHz
350	18 kHz	45 kHz	140 kHz
156	10 kHz	20 kHz	55 kHz

A full report in the 1993 World Radio TV Handbook when we test the final version of the Lowe PR-150. Designing a preselector is not easy at this price level, but the choice of varicap diodes as opposed to more expensive mechanical technology has its drawbacks. In a reaction to this review Lowe say they are working to improve the third order intercept point.

SUMMARY

In short, the Lowe HF-150 is an excellent choice as an entry level communications receiver. It gives much better performance than sets like the Trio Kenwood R1000 which were on the market 10 years ago for the same price, showing that it is still possible to improve on performance and keep the costs reasonable. Further information from: Lowe Electronics, Chesterfield Road, Matlock, Derbyshire, DE4 5LE England. Tel: +44 629 580800. Fax: +44 629 580020.

LOWE HF-225

In the first part of 1987, a new shortwave communications receiver came onto the European market. The Lowe HF-125, as it was called, has now been upgraded to the HF-225. Designed and built by Lowe Electronics of Derbyshire England, it is lightweight at 1.9 kg in its standard version, 2.6 kg if internal Ni-Cd batteries are installed. It measures 253 x 109 x 204 mm making it very compact. It is a double-conversion superhet design, with a high first intermediate frequency of 45 MHz. The set comes with a 12 volt DC power adapter for use on 220 volts AC current in Europe. This external power supply gives no problems, providing the transformer box is kept well away from the receiver (i.e. do not put it next to the receiver, or AC mains fields tend to couple with the receiver's oscillator circuitry).

The beginner will have no problems connecting up an antenna, a pair of headphones or a tape-recorder. Lowe electronics have written a clear, logical instruction manual plus a booklet that explains what to listen out for when you start tuning around. Part of its attraction is the simplicity. There are just 9 controls on the front panel, which can be tilted up to face the operator. Up to 30 frequencies can be stored in the radio's memory. Data is held in these memories using a lithium battery back-up. The memory channels can be quickly selected using the manual tuning knob. The tuning knob itself has a good feel to it, being large enough and offering a recess for the finger.

TUNING

A large clear green-colored back-lit liquid crystal display shows the frequency being received with a resolution nearest 1 kHz. Frequency display shows the true carrier frequency, whatever the mode selected. The manual tuning knob offers FAST and SLOW tuning, depending on the speed with which the knob is rotated. The tuning increments vary with mode selected as follows:

MODE	NORMAL SPEED	FAST SPEED
LSB, USB, CW	8 Hz	250 Hz
AM	50 Hz	500 Hz
FM	125 Hz	500 Hz
AM SYNC	8 Hz	NO FAST MODE AVAILABLE

The standard AM steps proved to be small enough for comfortable tuning without any annoying "chuffing" effects. If you select either upper/lower sideband or the CW mode as you would do for radio amateurs or telex stations, the slowest tuning step drops to 8 Hz. This is a considerable improvement over the older HF-125. Since the receiver is stable, the fine tuning steps allow you to accurately tune in a radio-teletype station with ease. The receiver also has a narrow-band FM option for monitoring citizen's band on 27 MHz for instance, but we didn't test this mode ourselves. The increment chosen for the AM sync mode is fine, but there is no provision for selecting either upper or lower sideband. We think this is a limitation. The keyboard comes as a separate option costing £39.50. It is a good idea to regard this option as essential, as tuning is made a lot easier. The keypad can be placed horizontally in front of the set, making it easy to punch in frequencies. The keyboard software is clever.

STABILITY AND BIRDIES

We put the receiver in a constant temperature environment (18 Centigrade) and tuned it to 12100 kHz. The set was then switched to the USB and LSB modes. After an hour, drift was less than 40 Hz in each instance. This is comparable with receivers in much higher price classes.

If you put a shortwave receiver inside a metal cage so that it's screened off from the outside world, you can then measure "ghost" signals on the dial generated by the set itself. They appear as a silent carrier if you're listening on AM or as a whistle if you're listening on single-sideband. Most sets of this price range suffer from these so-called "birdies", and you can't get rid of all of them without drastically increasing the price of the radio. The designer's aim is to make them as weak as possible and on frequencies that are outside major broadcast and amateur bands. We measured more than 25 low level birdies on the HF-225 (i.e. greater than 0.5 microvolt). Most of these disappear below the noise level when you connect an antenna, but we found 6 "birdies" which were above 1 microvolt, and therefore noticeable. If you try to tune in a station on these channels, you will get extra distortion or interference generated by the HF-225 itself. More expensive sets like the Yaesu FRG-8800 and ICOM IC-R71 have around the same number of birdies too. For a receiver of this price bracket, the number of birdies on the HF-225 is good, and better than the HF-125.

SENSITIVITY

In our tests we checked how much signal is needed at the 50 Ohm input to get a signal + noise/noise ratio of 10 dB. Sensitivity often varies with frequency, and for that reason the HF-225 was measured at several points in the spectrum. We checked the AM mode (using 60% modulation) and SSB. The 4 kHz filter was selected for the measurements, being the filter that SW listeners are most likely to use for broadcast listening.

Sensitivity: In microvolts at 50 Ohms: 10 dB S+N/N, 60% mod AM, 100% SSB, 1 kHz tone.

MHz	0.2	2.2	8.0	12.1	16.0	26.0
AM	0.66	0.63	0.70	0.73	0.68	0.71
SSB	0.22	0.20	0.20	0.19	0.18	0.28

The receiver's built-in synthesizer generates a lot of noise between 30 and 70 kHz, way down in the very low frequency part of the spectrum. This masks most of the signals coming in. But from 100 kHz up to 30,000 kHz, covered continuously on this set with no breaks, the sensitivity is remarkably uniform. Sensitivity is also dependent on the chosen bandwidth filter. The following measurements were taken at 12.1 MHz.

Sensitivity in microvolts, measured at 50 Ohms:
10 dB S+N/N against bandwidth AM 60% at 12.1 MHz.

IF (MHz)	2.5	4.0	7.0	10.0
SENSITIVITY	0.49	0.71	0.70	0.78

The receiver offers an analogue signal strength meter which turns out to be very accurately calibrated, far more than just a tuning gimmick.

RF ATTENUATOR/ AUTOMATIC GAIN CONTROL

A push-button attenuator is provided, offering around 17 dB attenuation when measured at 12 MHz. It consists of a resistance-network, which can be shorted out of circuit by a diode. After the initial attenuation is seen on the S-meter, there is no further indication that the attenuator is switched in.

The AGC is designed to keep the audio level at a fairly constant level, even though the signal may be fading. Using AM 60%, 1 kHz tone, at 12 MHz the audio level varies by 3 dB for antenna input between 0.7 microvolts and 120 millivolts. The speed of the AGC is not selectable, and we found the time constant chosen to be adequate.

STATIC SELECTIVITY

The HF-225 offers 4 different selectivity positions. There is a very wide 10 kHz setting, best suited when you're listening to local mediumwave stations, and then the low distortion audio (less than 3% at 100 mW) sounds very pleasant indeed to our European ears. The 7 kHz setting is best suited to shortwave when you've found a strong clear signal. The 4 kHz option is more useful when the signal you're listening to is suffering from strong stations operating on frequencies 5 kHz either side of the one you're trying to monitor. And finally, if you're trying to dig something out of the noise, or looking for utility stations, the 2.2 kHz filter is the best suited.

The first IF filter at 45 MHz has a fixed bandwidth of 15 kHz. In the second IF, use is made of switchable ceramic filters which are cascaded to improve the overall skirt selectivity, thus giving a 6:60 dB shape factor around 1:1.7. This is above average for the price range.

GOOD DYNAMIC RANGE

Dynamic range is a measure of how large the difference in signal strength has to be between a weak and strong signal, before interference by the strong signal occurs. This is important if the receiver is to be used for DXing. Again using the standard CEPT measurements, the two signal generators are tuned to 10000 kHz and 10020 kHz respectively, i.e. 20 kHz apart. For the HF-225, the dynamic range turns out to be 88 dB.

NO ACTIVE ANTENNAS

The Lowe instruction book warns against connecting huge antennas. We found an active antenna such as the Dressler ARA 60 was totally unsuitable with the HF-225 in Europe. With the 20 dB internal attenuator switched on, the active antennas were delivering signals than the radio could just about handle. The 10 meters of long wire or dipole recommended in the instruction book gave much better results. Lowe does market a special active antenna for use with this set, but we did not test this. In practice we found the mediumwave performance to be good. In January 1992 we re-tested the HF-225 with a mediumwave loop, and were able to log North American East Coast stations during good conditions.

PRICE

The HF-225 costs £429 including Value Added Tax in Britain. For an extra £39.50 you can plug in a keyboard, and then enter frequencies directly into the set. The software here is friendly, simple and logical. For another £39.50, you can also have a FM board installed, plus a synchronous AM detector. This mode generally offers better audio fidelity on shortwave signals, especially during deep fades. Although the synchronous detection system does perform, it doesn't lock onto very weak signals, and when this mode is used, the audio sounds less pleasant. That comment is very subjective though. But unlike the cheaper SONY ICF-2001D, which also offers synchronous detection, you cannot select between upper and lower sideband. We would rate the value of the synchronous detector as only FAIR.

FURTHER OPTIONS

You can also have a rechargeable battery supply, which is another £49.50. For better fidelity (especially on medium-wave), Lowe have introduced the S-225, this being a separate Wharfdale speaker measuring 190 x 203 x 240 mm. This costs £49.50 in the UK. A portable carrying case for the set, C-225 is also available at £23.86 including VAT.

CONCLUSIONS

The manufacturer's specifications check out perfectly, and Lowe have achieved all the aims they have set themselves. The manual supplied with the radio was particularly clear and concise. The HF-225 will be of interest to the serious shortwave listener and beginning DXer looking for a no-frills receiver which sounds good, and can handle a wide range of signal modes. The cabinet is simple but durable and difficult to scratch. This is important if the receiver is to be used for portable work. The dynamic range is good for a receiver of this type, the 3rd order intercept point for interfering transmitters more than 20 kHz from the desired frequency being 13 dBm...a very acceptable figure.

It is clear though that with the launch of the HF-150 (which has a better sync function) the HF-225 is due for an upgrade or a receiver above it.

Lowe tell us this will be in late 1992 or early 1993. It is hoped that the new receiver can be reviewed in the 1993 World Radio TV Handbook. In the meantime, the prototype shows that they are working on their design as well as the technology.

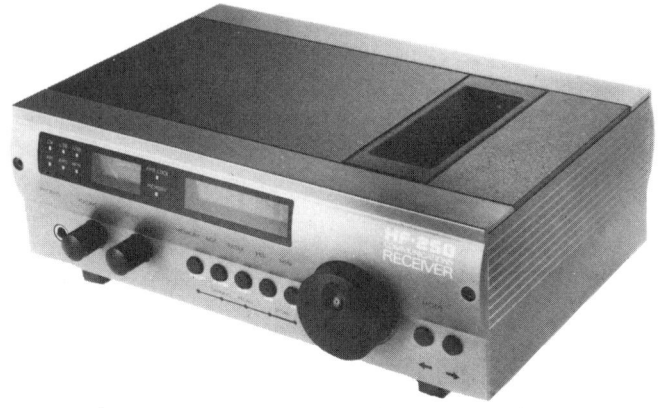

LOWE HF-250

The set is mainly being distributed in Europe and North America. Further information from: Lowe Electronics, Chesterfield Road, Matlock, Derbyshire, DE4 5LE England. Tel: +44 629 580800.

PANASONIC RF-B65L

This set is designed for the traveler, being about the size of a paperback book, and weighing in at 771 grams including the 6 penlight batteries that fit inside. That is certainly compact and light enough for someone on the move. When switched off, the digital liquid crystal display shows the local time, or you can set it to show a second time, e.g. UTC . So let's switch it on, by pushing the "ON" button at the top left hand corner.

At first, pushing the button does nothing, in fact pressing any of the keys on the front panel gives no response. That's not a problem, it's a built-in safety device.

On the side of the set is a stiff switch marked HOLD. When it's pushed to the ON position, the set remains inactive. This is to prevent the receiver being accidentally switched on by knocking against something else in a suitcase. If you slide the HOLD switch down, and then press the ON switch, the set springs to life picking up the last frequency it was tuned to. The set covers FM between 87.5 to 108 MHz, longwave between 155 and

519 kHz, mediumwave between 520 and 1610 kHz, and shortwave between 1615 and 29999 kHz. Some versions of the RF-B65 are being sold in parts of the world without longwave, and in some countries (e.g. Italy) shortwave coverage is deliberately restricted to an upper limit of 26100 kHz.

TUNING

To change to another station you have three methods of tuning the set. First there is a calculator style keyboard. 12 soft-touch keys are marked with the digits 0 to 9, one shows a decimal point, and the final button is marked M for memory. If you know the frequency you want to listen to, say 6020 kHz, you push the frequency entry button, then 6 0 2 0, and a button marked enter and there you are. If you tapped in 200 the set springs to 200 kHz longwave, if you put 98.95, it springs to 98.95 MHz VHF FM. Somebody has developed some very intelligent software. Each of the keyboard keys is also marked with a shortwave broadcast meter band. For example the "8" key is also marked with the 25 meters in small black letters. Push a key marked "meter", and then the number "8" key, and you jump to the lower end of 25 meter band, 11650 kHz to be exact. That's handy if you can't remember the frequencies where particular broadcast bands are located.

If you don't know the exact frequency a station is using, a keyboard isn't much use to you. You need to pick a part of the dial, and tune about a bit to find something of interest. The Panasonic RF-B65L offers two ways to do this. On the front panel are two buttons for moving either up or down the band. If you push the "UP" button, the receiver jumps up the band 5 kHz on shortwave, 100 kHz on FM, 9 kHz on longwave and a small switch in the battery compartment lets you choose either a 9 or 10 kHz step on mediumwave. That is important because stations in North America are 10 kHz apart, in Europe they are 9 kHz apart, and the UP and DOWN buttons are clearly designed to let you switch to the next station along the dial. You can also automatically scan a chosen meter band for strong listenable signals.

CONVENTIONAL TUNING KNOB

In addition, there is a conventional rotary tuning knob. It can be set to make the set jump up and down the bands, in the same way as the push buttons just described. Or you can set it to tune more finely, in 50 kHz steps on FM, and 1 kHz at a time on short, medium, and longwave. This allows very accurate tuning of the set for broadcast listening. In addition you can store the frequencies of 9 favorite shortwave, 9 favorite mediumwave, 9 favorite longwave, and 9 favorite FM stations in the 36 channel memory. Signal strength is shown on a rather neat colorful electronic "S" meter under the frequency display, though the meter tends to overestimate a signal's rating...everything appearing strong on some lower parts of the shortwave dial.

This radio is purely for listening to broadcast stations. There is no single-sideband facility to allow you to listen to amateur radio operators, radioteletype stations, or Morse code signals. The single 5.5 kHz bandwidth filter means the radio gives good fidelity on all bands, though if two strong stations are operating 5 kHz apart, you may have to slightly off-tune to separate the signals from each other. Picking out a weak station wedged between two strong stations 10 kHz apart is difficult. An optional narrow filter might have been handy, though it would add to the price. Sensitivity of the receiver is average for a set of this type. Under normal conditions the adjustable telescopic antenna will be sufficient, though a provision is made for connecting a short external antenna. A switchable signal attenuator ensures that the set doesn't overload on the 49 meter band at night when used in Europe. A standby switch allows you to switch the set on at a preset time for a fixed period of 90 minutes.

COMPETITOR

In terms of size, weight, and facilities this set is clearly designed to compete with the SONY ICF-7600DS portable radio, also known as the ICF-2003. Sony's radio has been around for some time. The new dual-conversion Panasonic model is virtually identical to the Sony when it comes to sensitivity, and selectivity, that's the ability of the set to pick out the station you want from the interference. The audio output power on the Panasonic is noticeably louder, even though in terms of power 550 against 400 milliwatts on the ICF-7600DS is not much extra. Not only is the loudspeaker on the RF-B65L more efficient, it gives a much richer sound than the rather tinny audio on the Sony set.

The set consumes a current of about 70 mA at normal listening volume. That quite a drain on the 4 penlight batteries in the radio section. Listening 3 hours a day, the batteries won't last much longer than about week. An optional 6 volt DC power supply is available to allow you to use the set off the household current supply. We used rechargeable nickel cadmium batteries with some success, though because these give slightly lower voltage, the set is slightly less sensitive on higher shortwave frequencies as a result. Rechargeable batteries last a shorter time than ordinary dry cells, but can be used again of course. The radio shuts itself off when the battery voltage falls below a certain level.

PRICE AND CONCLUSIONS

The receiver costs US$280 in North America, or £179 (DM540) in Europe. If you're interested in listening to broadcast stations only, and want the simplified tuning that this set provides, this receiver gives good value for the technology it offers. It is NOT a set for the hobbyist interested in other types of signals on the shortwave bands, but then it's not designed to be. The operating instructions with the set are clear and comprehensive. However, the free guide to shortwave station frequencies is several years out of date, rendering it useless. The Panasonic brand is sold by electronic stores in most countries of the world.

PANASONIC RF-B45

Panasonic have not been leaders in the shortwave portable business, usually following trends blazed by other Far Eastern companies. Now, however, it appears they have really thought about their place in the market and quietly introduced an excellent piece of work.

Launched world-wide in early 1991, this dual-conversion receiver is clearly based on its predecessor, the RF-B40, with improved fine-tuning and the additional capability of single-sideband reception. But in the process of upgrading the receiver, Panasonic has added some of the features seen on the RF-B65L and build it into a smaller case. It has an illuminated on-off button so you can see at a glance if the radio has been left on with the volume turned down. On most versions, the set covers longwave between 146 and 288 kHz, mediumwave between 522 and 1611 kHz, and provides continuous shortwave coverage between 1611 and 29995 kHz. FM between 87.5 and 108 MHz is also offered. On models sold in Italy the coverage of the receiver is partly restricted in the shortwave range.

The RF-B45 can be tuned in one of two ways. On the front is a block of 12 black keys, similar in layout to a pocket calculator. Direct entry of the frequency is possible after you press the "FREQ" key. 10 of the push buttons on the keypad also have a small text on them, corresponding to 10 different broadcast bands between 3 and 22 MHz. If you are on your travels and can't recall where the 25 meter band starts for instance, you simply push a button marked meterband. Then, by pressing key number 6 on the panel, the set springs to 11650, i.e. the lower frequency end of the 25 meter broadcast band. Manual tuning is offered too, via two well-designed push buttons in the top right hand corner of the set. On shortwave, pressing one of the buttons moves you up or down the dial in 5 kHz steps. If you keep the button pressed, the receiver continues to move up the dial at about 50 kHz a second. On mediumwave you can select whether the receiver moves either 9 or 10 kHz at a time. On longwave the steps are 9 kHz, and 50 kHz on FM. But note that the set "chuffs" as it scans.

Fortunately, unlike the old RF-B40, the new RF-B45 model has a fine-tuning control. This takes the form of a thumbwheel on the side of the receiver which is adequate for most broadcast listening. Resolving SSB signals is more difficult, but it does not take long to get the hang of things. The set offers 18 memories (9 on AM), rather than the 36 offered on the RF-B65. If you press the "0" button without the FREQ button, the radio scans the contents of the 9 memory channels in that mode. That is useful if you program the channels with alternative frequencies of your favorite stations.

Performance on shortwave is fine in Western Europe. In more remote locations, a short piece of wire may be needed to improve the strength of higher frequency shortwave signals. The two-step attenuator helps reduce the effects of nearby medium-wave stations. There is only one bandwidth filter which is only fair for a portable, being rather too wide for shortwave broadcast listening. The problems can be somewhat offset by carefully adjusting the fine-tune control. The radio can be set to come on at a predetermined time, but there is no remote outlet to drive a tape-recorder. The radio also has a sleep function which automatically switches off the radio after 90, 60 or 30 minutes. Sadly, the radio only displays the time when it's switched off...old habits die hard.

The radio uses 4 penlight cells, drawing a steady 50 mA at comfortable listening levels. In some countries the radio comes complete with a power supply. The "bar style" signal strength meter looks impressive at first, until you realize that there are really only three positions. You can't really use this as a serious tuning indicator, except maybe on FM. FM reception is in mono, and you'll need a mono-stereo adapter if you want to plug in stereo headphones, otherwise you'll only get sound in the left ear. The tone control is simply a two-position switch, and unfortunately there is no dial-light.

At a price of around US$170.00 this set offers strong competition to the Sony ICF-SW7600. In Europe the price is around US$220, making it considerably cheaper than the Sony alternative. It also offers similar features to Panasonic's other serious portable, the RF-B65. In short, Panasonic have come up with a excellent receiver for the traveller.

PHILIPS (MAGNAVOX) AE-3805/ SANGEAN ATS-800

Back in the late 1920's and up until well into the sixties, the Dutch Philips Company enjoyed an excellent reputation in the field of domestic valve radios. Models were made by Philips at factories in Holland and Britain, the more expensive radios making use of the pentode valve which was a Philips patent. After all, making radio valves or tubes wasn't far removed from the techniques used to make light bulbs.

Many of the Philips domestic radios offered excellent shortwave coverage, often with a band spread control to separate the stations in the crowded shortwave bands. Alongside radios for use in the home, Philips also built up a name in the professional communications receiver world.

At the start of the 90's Philips in Europe appeared to have said good-bye to the shortwave portable world. Three models were on offer, a compact portable called the D1875, a larger table top portable called the D2935 and the serious hobby receiver, the D2999. The radios have been available in North America under the Magnavox brand. Philips continues to manufacture shortwave portables in India, but these are only for the domestic market, and bear little resemblance to radios from the same company sold in other regions of the globe.

At the start of the 1990's Philips announced two new shortwave travel portables, and at the same time discontinued the larger portable radio, the D2935 and the hobby receiver, the D2999. The two new radios indicate that Philips is only interested with the business traveler market as far as international radio is concerned. We purchased the more expensive AE-3805 for review.

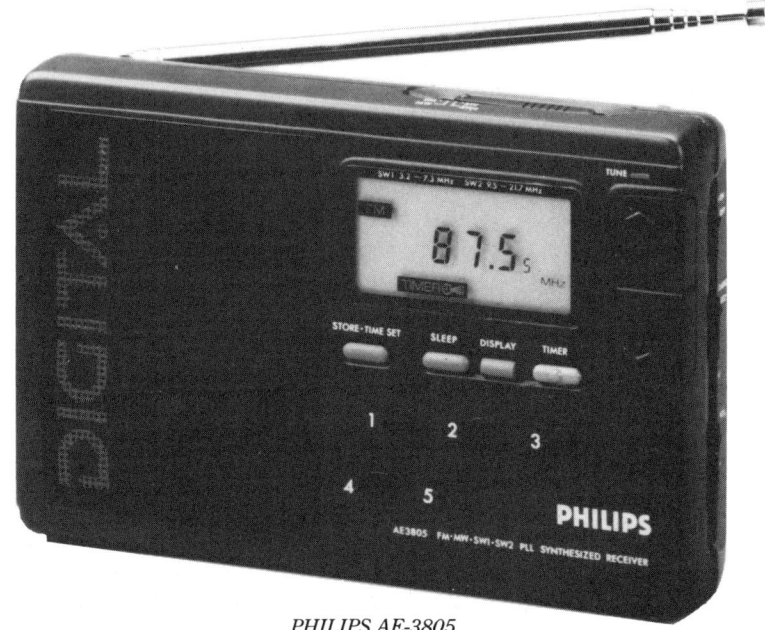

PHILIPS AE-3805

Launched in 1990, this set is manufactured for Philips by Sangean of Taiwan. Sangean sell a similar receiver under their own brand name, calling it the ATS-800. Siemens of Germany also carry the radio as the RP-647G4. In North America, Tandy Radio Shack stores carry the same receiver at the DX-370. There are some cosmetic changes to the front panel, and FM stereo (via headphones) is available on some versions. Performance on medium and shortwave is identical to the Philips AE-3805!

The radio measures 17 x 11 x 3 cm, which means you can grasp it firmly in the palm of your hand. It weighs 504 grams including 4 penlight cells which clip into the back....all in all a compact portable. The front of the set has a liquid crystal display, a row of small buttons to adjust the built in clock and timer, plus a set of buttons marked one to five. In fact these buttons are not for direct frequency entry, which is impossible, but they are used to store favorite frequencies in the memory. You get 5 on mediumwave, 5 on FM, and a total of 10 on shortwave. It is very easy to set frequencies in this memory system.

COVERAGE

There's no longwave on this set, so if you rely on that band for listening to stations in Ireland, Britain, France or Germany for instance then you need to make another choice. On mediumwave, the European version skips up and down the dial in 9 kHz steps. However if you take the radio to North America, there is no provision to switch to the 10 kHz spacing used by mediumwave stations there. The reverse is true of the Magnavox model sold in North America. That means that if you use the European Philips 3805 in North America there would be some local mediumwave channels that would be so off tuned that reception would be distorted. FM coverage is standard 87.5 - 108 MHz in mono.

COMPLICATED TUNING AND INCOMPLETE HF COVERAGE

Shortwave coverage is divided into two bands. SW1 corresponds to 3.2 - 7.3 MHz, and the other SW2 works out at 9.5 - 21.75 MHz. That means that you can't pick up quite a lot of the 41 meter band frequencies used by private stations in the US, or some 31 meter band channels used by the BBC and BRT in Belgium.

A very brief and poorly written instruction book explains that you select a frequency by using the up and down buttons on the side of the set. So, for instance, you start at the bottom of the frequency range and step up in frequency until you reach the desired channel. Then you might want to store that in the radio's memory. The set scans the band in 5 kHz steps on shortwave, beeping as it does so. The designers have compromised on the cost of the display by making the last digit smaller in size and only able to show a 5. So, if you select 11855 kHz for instance, the display shows 11.85 in large numbers, followed by a smaller 5. That might be rather confusing to someone not used to shortwave frequencies. It's compounded by the fact that 11860 kHz is shown as 11.86 in the display, with no trailing zero. There is no dial light to see the display in a darkened room.

The radio is extremely sensitive on shortwave. On mediumwave the sensitivity is deliberately reduced to prevent local stations overloading the receiver's delicate input circuits. On paper, the radio also appears selective for its price. Philips have chosen a sharp single bandwidth filter which means that the radio can easily separate international broadcasters just 5 kHz apart.

DREADFUL DYNAMIC RANGE

This radio really falls down badly in the field of dynamic range, or the set's ability to pick up weak signals in the presence of strong ones further down the dial. Reception on the 49 meter band is reduced to the stronger stations only....in fact in the evenings in Europe, the best results are obtained when the telescopic antenna is fully collapsed.

The set's single conversion design also means that image rejection in one of the worst we've ever experienced on a digital portable. You not only get signals on the frequency the station announces, but also 900 kHz below the proper channel. Maybe it is fun to have a radio that picks up Radio Netherlands on 12800 kHz but that's not the intention. Besides, it also means that utility stations around 10.5 MHz break through and cause unwanted whistles and Morse interference on the 31 meter band. The performance after dark can only be described as chaotic. We tested two off-the shelf samples in 1990, and a third sample in August 1992. One of them was totally unstable on the 60 and 90 meter bands, being unable to handle the incoming signals and just going into oscillation. The radio comes with a guide to shortwave frequencies which unfortunately is a list of channels of the past. It is so out of date as to be seriously misleading.

CONCLUSIONS: LOOK ELSEWHERE!

In short the Philips AE-3805 is a disappointment. It costs around US$150 in Europe and around US$100 in the United States. For that price you can buy a good quality analogue portable with dual conversion circuitry and much better performance. The advantage of digital frequency readout on this radio is spoiled by the clumsy method of tuning. The limited frequency coverage and the inability to select between 9 and 10 kHz spacing on mediumwave all add up to an overall rating of fair-to-poor. These conclusions also apply to a certain extent to the Sangean ATS-800 and Radio Shack DX-370, and the Siemens RP-647G4. But the price of these sets is often much lower than the Philips version. The Siemens RP-647G4 is currently selling in Europe for around US$87.00. This makes much better value, and for this price maybe you can live with the imperfections. But can't Sangean make a PLL digital receiver with better image rejection and sell it for US$95?

PHILIPS AE-3905

In 1991 Philips celebrated 100 years of its existence. This anniversary came at a time when the company faced some serious budgetary problems in its consumer electronics branch. At presstime that sector of Philips is still facing problems. Clearly as part of a scheme to show that it was not beaten yet, the winter 1991/2 product catalogue presented a number of unique products with an exclusive design. One of these products turned out to be an extremely compact digital world receiver, about the size of a cigarette packet. It is called the AE-3905, and they're right to claim that it's unique.

THE MAP DISPLAY

Half of the front of the receiver contains a LCD screen showing a world map, and two clocks. One shows hours, minutes and seconds which you

set to your local time, the display on the other depends on which time zone you set. Several major world cities such as London, Dakar, Hong Kong, New York, & Los Angeles are shown on the map, although curiously for Philips, Amsterdam failed to score. As you select the cities one of the clocks shows you the winter time in that city. Actually the front of the radio turns out to be a door. If you open it, on the back of the door is a keyboard for entering AM or FM frequencies, the AM coverage being between 147 and 29995 kHz.

TUNING STEPS

The steps on mediumwave are switchable between 9 & 10 kHz, and on shortwave the steps are 5 kHz with no fine tuning. Memories are also included. The other half of this miniature radio contains a 4 cm speaker, with the option of connecting a pair of headphones for stereo reception on FM. The tone control is switchable between a music and speech position. A very small lock switch on the side of the set which requires dainty fingernails to operate locks all the functions except the on-off switch which we find strange because it's so easy to accidentally switch on the radio when its sitting in a shirt pocket. Incidentally it is the first shortwave radio we know of that has been made in Austria.

PERFORMANCE

At a price in Europe around US$350, you clearly pay for cramming the technology into a small space. This is clearly Philips' answer to the Sony ICF-SW1, but with a LCD map attached. Compared to other sets in the same price range, shortwave reception is seriously disappointing. Despite the dual conversion design, image-rejection is poor, and the tiny speaker doesn't produce much volume. The lack of fine tuning in between the 5 kHz steps makes reception of weak stations on the crowded bands somewhat frustrating. This radio is clearly destined to become a collectors item, but it is not setting new standards as the advertising campaign would have you believe. It is already becoming difficult to find in Europe and appears never to have been launched in North America.

SIEMENS RK710

Launched in August 1991, this receiver is sold in Europe by Siemens, although the radio is actually made by Sangean of Taiwan. Siemens say that all receivers imported by them into Germany receive an extra level of quality control checks before distribution. At first glance, the liquid crystal display looks like it might be a frequency readout. However, the display serves to indicate the time, and the band which has been selected (e.g. SW8, 13 meters). The display can be illuminated at night. A symbol also becomes visible when a strong station is received....but your ears are a much better judge of correct tuning than this crude form of strong-signal indicator. Another element of the display is used to indicate stereo reception on FM. The set has two built-in clocks so, for example, one can be set to local time, the other to UTC.

However, the size of the liquid crystal display means that the analogue dial has been squashed as result. Although the major international shortwave broadcast bands are available (i.e. 49, 41, 31, 25, 22, 19, 16, & 13 meters) each band is squashed into a dial length of 3 cm. This makes the band spread very poor. Note that some out-of-band channels in the 31 meter band are not receivable on this receiver (e.g. 9410 kHz) since coverage there is between 9450 and 10000 kHz. Coverage on the other broadcast bands is fine.

Performance of this single-conversion portable is not helped by very poor image rejection. Even on quieter shortwave bands (e.g. 21 MHz during daylight hours), mixing products from utility stations outside the broadcast bands appear all over the dial. This causes annoying whistles and Morse-code interference to many broadcast signals, giving the first-time user the false impression that shortwave broadcast bands are more crowded than they really are. The built-in 5 cm speaker gives a somewhat "thin" sound, so headphone listening for FM is recommended. The set is supplied with "in-the-ear" stereo headphones.

The set also offers standard mediumwave (AM) reception, plus FM in stereo via headphones. No longwave is offered. Battery consumption (47 mA at normal listening levels) is average for a radio of this type. It weighs 398 grams with the three AA-size batteries installed.

The price in Europe is US$80, which makes it somewhat expensive bearing in mind the only mediocre shortwave performance. Siemens has recently upgraded the set to the RK713 which includes a world-time function. It has the time of less than 285 cities programmed in to it. However since the radio has no date function it cannot keep track of summer and winter time changes. Shortwave performance is a little better than the RK710, but nothing special.

SIEMENS RK631

This 10-band set is halfway between the pocket-book sized portables, and the "cigarette-packet" size of shortwave radios. Apart from mediumwave (AM), longwave, and FM (in mono), the receiver offers coverage of the 49, 41, 31, 25, 22, 19 and 16 meter shortwave broadcast bands. The dial offers enough coverage of "out-of-band" channels, these frequency ranges being depicted in a different color. The set weighs 390 grams when the two AA-size batteries are fitted.

Despite the relative ease of tuning thanks to the band spread, performance on shortwave is only fair at best. The image rejection circuitry of this single-conversion set is poor, leading to a lot of unwanted whistles and utility interference. The lack of the 13 meter band is a problem for listeners planning to travel to or live in Asia....many international broadcasters make extensive use of that band. Power consumption, at 45 mA for normal listening levels, is not excessive. The 3.5 centimeter speaker provides communications quality audio, but this is not a disadvantage when listening to speech programs on shortwave.

The price is around US$45 in most European countries. This radio is only really suitable for listening to the stronger stations, and even then, the chances of unwanted mixing products spoiling reception is rather high.

SIEMENS RK670/ SANGEAN ATS818(CS)/ TANDY DX-390

In previous test reports we've remarked that compact radio-cassette recorders with shortwave are usually a poor compromise. You're better off buying a good shortwave radio and connecting an external tape recorder. In the past two years two cassette recorders with PLL synthesized shortwave radios have appeared. Only one is worth a full review.

This radio has been made for Siemens by the Taiwanese Sangean company. Sangean also sells the radio under its own brand name (the ATS 818CS) but it doesn't have the factory pre-programmed stations in it. Radio Shack in North America have also launched a set called the DX-390 which is similar, except that the cassette recorder mechanism is not included. As a result, a larger speaker is used. The British company of Roberts Radio in West Molesey Surrey also import the radio, marketing it as the RC818 and putting a price of £200 on it.

The Siemens RK670 is designed as a table-top receiver. You can either operate the set upright, which makes the buttons for the cassette recorder easy to operate, or unclasp a stand on the rear of the set. This lets you tilt

Siemens RK670

the front panel at 45 degrees, making the radio operation much easier. A total of seven batteries are needed. Three penlights are used to run the clock and computer section of the radio. An additional four "D" size cells are needed to power the radio itself. Power consumption at an average listening volume is 80 mA. Switch on the cassette recorder and the consumption rises to 150 mA. If the radio is playing, and recording to the cassette at the same time, power consumption is in the region of 200 mA. We obtained a battery of life of just 4 hours with cassette and radio playing simultaneously, before the four alkaline "D" cells needed replacing. There is a noticeable increase in distortion, and a letter "E" is displayed. So it makes sound economic sense to use the 6 V DC mains adapter that is supplied, at least with the Siemens version of the radio.

TUNING

The radio has a keypad for direct frequency entry. You push a button marked frequency and tap in the desired channel. The software is clever..."9890" is recognized as a shortwave channel, where as "98.90" is regarded as FM. There's a simple button sequence to find the lower end of each of the shortwave bands.

Manual tuning is achieved in two ways. "UP" and "DOWN" buttons allow you "jump" to the next channel. On shortwave the steps are 5 kHz apart. On medium/longwave the steps are 9 kHz apart, but unfortunately the steps do not correspond with the new longwave plan for Europe. The set jumps to 200 kHz for BBC Radio 4, when in fact the station is on 198 kHz. If you depress either of the UP-DOWN buttons for more than a few seconds the radio starts to scan, stopping when it hits a strong signal.

A tuning control on the side of the set can either be set to do exactly the same as the "UP-DOWN" buttons, or set to a "FINE" position. This allows tuning in 1 kHz steps on LW, MW, and SW, 50 kHz on FM. The receiver can receive single-sideband signals thanks to a beat frequency oscillator. A small rotary knob is used to clarify the signal....not as easy as with sets that have USB or LSB selection.

The Siemens version of the radio comes with 18 radio stations pre-programmed at the factory. For the most part these are stations which have German language programs, not surprising seeing that the main market for this set is in the German speaking parts of Europe.

SENSITIVITY AND SELECTIVITY

The RK670 is sensitive enough on all the bands, although in low signal strength areas, a piece of long wire clipped to the antenna may be needed for the reception of signals higher than 15 MHz. There is also a connection for an external antenna, although long wire antennas above 10 meters simply overload the radio. The receiver offers two bandwidth filters (6/2.8 kHz @ -6 dB) for AM/SSB reception. The shape factor of the filters is fair. Image rejection on mediumwave was found to be mediocre, and on shortwave it is also only fair. The RF gain control didn't seem to have much effect on reducing spurious signals, except to demand that you turn up the volume. The dynamic range of the radio is fair-good compared with other sets in this price bracket and better than the Sangean ATS-803A.

AUDIO

The inclusion of the cassette recorder means the speaker in the radio has been reduced in size. In fact there are two, one 2 cm tweeter and a 7 cm woofer. In general we found the audio to be somewhat muffled, this being especially noticeable when headphones are connected. Reception on FM, and cassette playback, is in stereo when using headphones.

CASSETTE RECORDER

The stereo cassette recorder has a tape-type selector switch but no Dolby. The timer in the radio can be used to switch the recorder on at a pre-determined time to tape a program unattended. The set has a built-in microphone, but you can only use it for logging notes. Sets sold in late 1991 had problems that the background hum from the drive mechanism ruled out any serious recordings. We retested the radio in 1992 to find that the hum problem had been reduced, although the microphone quality is really designed for note-taking.

CONCLUSIONS

The Siemens RK670 retails at around US$300 in Europe. The addition of a cassette recorder has compromised the sound quality of the radio, but the price is attractive for those who want an "all-in-one" unit. It is more than adequate for general program listening, although the image rejection may create problems for those interested in digging the weaker stations out of the noise. We rate it as good value. In the US, Radio Shack sell the DX-390 for US$240, which is somewhat expensive bearing in mind the features and performance are almost identical to the DX-440 which retails at US$40 cheaper. The DX-390 has no cassette recorder. Sangean appears to sell the ATS-818CS under its own brand name in the US as well (e.g. via EEB, Vienna Virginia). The price for the unit with a cassette recorder is US$270, and US$220 without the cassette option.

SANGEAN ATS-808
(SAME AS SIEMENS RK661/ROBERTS R808/ AIWA WRD1000/RADIO SHACK DX380)

SIEMENS RK661

This radio was originally launched in the middle of 1990, With a price tag in Europe around £120, this set is designed to heavily compete with other "paperback book" size digital portables. The ATS-808 measures just 197 x 125 x 37 mm, and weighs 695 grams including the 6 penlight batteries (two of these are used to drive the on-board computer, the other four to run the radio). It will be marketed alongside the older (and much larger) ATS-803A for the duration of 1993 at least. In the US the ATS-808 retails for around US$200. Do not confuse it with the ATS-800, a very different receiver. The British company of Roberts Radio in West Molesey Surrey also import the radio, marketing it as the R808 and putting a price of £120 on it. In the USA, Radio Shack sell the same radio as the DX-380 and recent catalogues show a good price for it..US$180.

The ATS-808 is designed for simple operation. It is for the reception of broadcast signals only. Although it has a stable PLL synthesizer, there is no provision for SSB (unlike on the more expensive Sony ICF-SW7600 or Panasonic RF-B65). However, the ATS-808 has provision for two bandwidth filters, (6 and 2.7 kHz wide (-6dB)), both with a fair shape factor for a receiver of this type. Frequencies are shown on a LCD display which is easy to see in daylight. The memory channel number is shown in one corner of the display, the time (using 24 hr clock format) and the relevant meter-band elsewhere in the window. A simple "bar" style signal strength meter is also offered, although this was far too sensitive to be anything more than a rough indicator. At night you can't see any of this information because there is no back lighting of the display. The radio has a timer, and a dual clock... that is useful if you want to keep track of the difference between UTC and local time.

COVERAGE

The ATS-808 covers 150-1620 kHz, 2300 -26100 kHz, and 87.5-108 MHz. If you listen to the radio using headphones, then stereo reproduction on FM is possible. The tuning software has been well thought out. You can punch in the frequency on the keyboard. The keys are well spaced apart so that pushing two keys together by accident is almost impossible. However, unlike some radios which use a small raised dot on one key to help a visually disabled operator orientate his or herself, all the keys on the Sangean feel the same. You can also ask the radio to select a particular meter band if you cannot remember a particular frequency. The ATS-808 offers two means of manual tuning control, either a set of up-down buttons, or a manual rotary tuning knob at the side of the set. Depending on a switch setting, the tuning knob adjusts the tuning frequency by 1 kHz on AM (50 kHz on FM) or by 9/10 kHz on AM (100 kHz on FM). In Europe the radio should be set to jump in 9 kHz in the fast position, but this can be adjusted to the 10 kHz channel spacing. You can also lock the tuning system once you have found the channel you are looking for. Simple automatic bandscanning is also offered on this receiver.

PRE-PROGRAMMED MEMORIES

The radio offers 45 memory channels, nine each on LW, MW, and FM, 18 on shortwave. The first bank of 9 is selected by pressing a number on the keypad. The next 9 are selected by pushing the "0" button first, then the chosen number. In Europe, Siemens of Germany distribute the Sangean radio under their own brand label (i.e. RK-661). Siemens have had 18 channels pre-programmed in the factory, all of them popular frequencies for German language programs (e.g. Radio Austria International on 6155 kHz). A table on the back of the set tells you which channel represents which station and frequency. Some of the frequencies chosen are peculiar, but you can re-program some or all of them yourself. Another version of the same set is on sale as the AIWA WRD1000. Apart from cosmetic changes to the outside of the front panel, the performance of the radio appears identical. The AIWA has no pre-programmed channels. The radio works best on its built-in telescopic whip. The dynamic range of the radio is fair, overloading during evening hours being reduced to acceptable levels by reducing the length of whip or switching the attenuator to the LOCAL position.

In short, the ATS-808 is an excellent portable radio for the price. It is not a DX machine, and yet the two bandwidth possibilities make it more versatile than most portables in its price range.

SONY SW1S/E PORTABLE RECEIVER

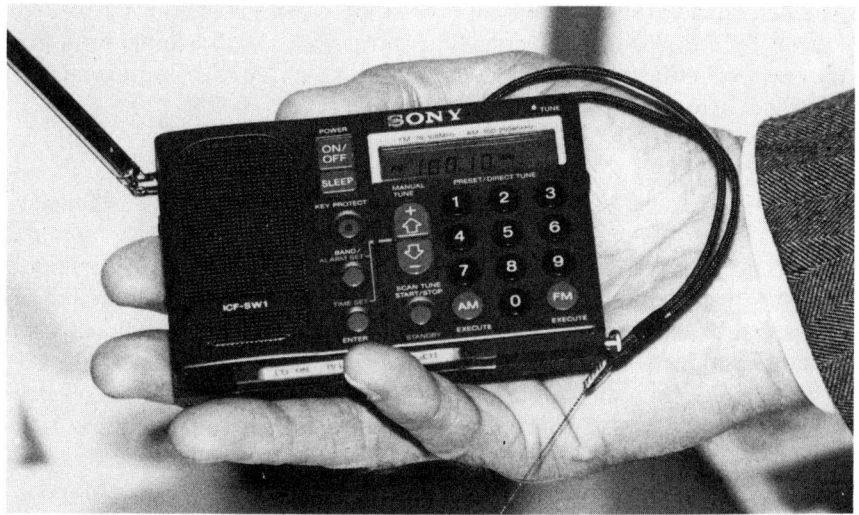

The "vital statistics" of this receiver work out at 118 x 71 x 24 mm. That is about the same size as most plastic cases round an audio compact cassette. The SW1 is just very slightly thicker. It will certainly fit in the breast pocket of a medium size man's shirt, and indeed the whole radio is designed around the need for compact electronics. It is designed for broadcast reception only, there being no provision for single sideband or CW reception.

The SW1 is sold in slightly different versions depending on where you buy it. In the example we purchased in The Netherlands, the set is tunable continuously between 150 kHz and 29995 kHz on AM, and between 76 and 108 MHz on FM. But in some countries like Saudi Arabia and Malaysia there is no coverage on the radio between 285 and 531 kHz. Sets bought in the Federal Republic of Germany don't cover above 26100 kHz.

OPERATION

To get it going you have to put 2 penlight cells into the back of the receiver. They slide in fairly easily, although you need to be careful not to force the batteries. Power consumption at a comfortable listening volume is in the region of 60 mA, which is quite economical for a computer controlled radio of this type. There is a small socket on the side of the receiver which allows you to connect a 3 volt DC power supply, which is supplied as part of the price in this part of the world. We noted the best battery performance when using nickel-cadmium rechargeable cells. Sony also make a special cable incorporating a resistance bridge so you could plug the set into the cigarette lighter of a car.

On the top of the set is a main power switch. Although this is deliberately quite stiff to prevent it switching on accidentally while in transit, it is not recessed. If you put the SW1 inside a suitcase against one of the walls there is a small chance the set would spring to life when jolted. But the risk is reduced by the fact that moving the main power switch simply puts the radio into standby mode. You then have to push another button on the front panel before the radio springs to life.

TUNING

There are two ways to tune the receiver. Suppose you know a frequency you want to select, e.g. 9770 kHz. You press a button marked AM, then 9 7 7 0 and then the AM button again to execute the command. A clear liquid frequency display indicates the chosen channel, and a button on the top of the set can be used to illuminate this display if you wanted to check it in the dark. The software is fairly user-friendly. However, if you select a frequency on AM, then move to FM, to return to AM you need to punch in a frequency again. Pressing the AM mode button is not sufficient. The lack of a fine tuning control is the most serious limitation on this radio. On shortwave, you move along the dial in fixed 5 kHz steps. If you try to tap in 9778 kHz let us say, then the receiver rounds that down to 9775 kHz as soon as you press the fourth digit on the keypad. There is no manual rotary tuning knob, but you can jump up and down the shortwave section of the band in 5 kHz steps using an "up-down" push button combination. The receiver "chuffs" along as you do this.

You can also push a button marked "SCAN TUNE" whereupon the radio moves up in frequency until it hits upon a fairly strong signal. It then stops, lets you hear the station for a second, and then continues scanning again. You can interrupt this process when the radio finds something you like. You can also store up to 10 favorite AM or FM frequencies in the radio's memory. Unlike some other portable computer controlled receivers, if you change the batteries within about 10 minutes, the information you have already programmed remains intact. No extra batteries are needed to run the computer section of the receiver.

FINE TUNING DRAWBACK

There is a small red light emitting diode designed as a tuning guide. But this is only really for effect on long, medium, or shortwave, especially in Europe or Asia where the dial is full of stations. If you are listening to a signal on 5995 kHz, and a powerful station starts splattering away on the adjacent channel of 6000 kHz there is no way that you can slightly off tune the radio to say 5994 kHz and escape the problem. This is possible on the cheaper (but larger) Sony receiver called the ICF-7600DS/ICF-2003 and competing receivers such as the Panasonic RF-B65.

Clearly Sony felt this was not necessary on this small set, and they have thus made a quite serious design error. Being able to fine tune is not a luxury in these days of overcrowded international radio bands, and there are even some stations in Asia that do not broadcast exactly on the 5 kHz channel, either deliberately or because of a technical problem. On medium and longwave the tuning steps are either 9 or 10 kHz depending on how you set a switch in the battery compartment.

GOOD SELECTIVITY AND DYNAMIC RANGE

Other specifications are good to excellent for a shortwave portable. There is a single 5 kHz bandwidth filter (- 6dB) which offers fair to good selectivity. The dynamic range of the receiver, especially on mediumwave, is a clear improvement over the ICF-7600DS model, and even better than the ICF-2001D. This is partly achieved by reducing the receiver's sensitivity. Despite the very small internal ferrite rod antenna provided, the mediumwave band is full of readable signals in Europe. Should overloading occur this is further reduced by switching a two step attenuator to the local position. Using the same receiver in North America we found that outside city areas, the lack of sensitivity required the use of the active antenna for mediumwave reception.

On shortwave you have to extend the small telescopic antenna for best reception. Overloading on the lower shortwave bands of 41 and 49 meters was only minimal during European trials. Sensitivity on FM has also clearly been reduced in comparison with the oversensitive ICF-2001D. This is fine for reception in cities where the transmitter is very close by. In country areas you might find the stereo reception on headphones somewhat noisy even when the telescopic whip is fully extended. Note that FM reception is determined by the angle of the telescopic whip ....on many of Sony's other personal AM/FM radios the headphone lead is made part of the antenna circuitry. Not in this case.

LOW TAPE RECORDER OUTPUT

The volume control comes in the form of a thumb wheel at the side of the set. It is very easily knocked, especially when the radio has been put into a pocket. As a result, the receiver can sometimes blast into overload as it is switched on. The audio output level of the receiver is adequate for indoor listening. The radio has a provision to switch itself on at a chosen time, so you can use it as an alarm clock. Once this is activated the set remains on for 65 minutes or until you otherwise interfere with it. When switched off, the frequency display shows the time providing the batteries have been left in. There is a stereo output socket for connecting to a tape-recorder, though Sony correctly point out that the level is extremely low. It is designed for portable dictation type recorders that only have a microphone input. If you are connecting a standard tape recorder, set the volume control around position 4, and plug the jack into the headphone output. A backlight for the LCD digital display is useful, and essential for tuning the receiver in the dark.

ACCESSORY KIT

In some countries you still cannot buy the SW1 mini portable receiver on its own. It comes in a plastic presentation case complete with several accessories. There is a smart power supply which you can plug into any wall socket regardless of the AC power. If it is between 100 and 240 volts it will work. In some countries like Britain and Holland Sony remove the power plug adapter from the package so you'd have to use your own travel plug adapter if you want to cross the Atlantic and use the radio in the US for instance. National safety regulations seem to be the reason for that move. Curiously, the plastic presentation case has no handle. You are supposed to attach a shoulder strap and carry it around your neck.

Inside the lid of the plastic case is a compartment containing a well written instruction book. There is also a frequency chart. Our copy was dated July 1991 which has limited use now, but at least the station address list is reasonably accurate. Which leaves us with two small boxes which together make up an active antenna. A small controller clips onto the side of the radio, and makes an inductive connection with the radio's retracted telescopic antenna.

The instructions tell you to extend the telescopic antenna on the larger black box and place this so-called antenna module near a window. There is a cable coiled up rather like a travel clothes line and you unwind this and plug the end into the antenna controller clipped onto the SW1 radio.

ANTENNA PERFORMANCE

If you are trying to listen in a concrete apartment block or hotel which completely shields out shortwave signals, then it might be an idea to try the supplied active antenna AN-101. Our tests show that the efficiency of the AN-101 is deliberately kept low. Not much of the signal picked up by the telescopic whip is delivered to the built-in FET amplifier. This means that intermodulation products are low, but on the other hand weak distant signals are buried in the noise generated by the antenna itself. Although suitable for listening to stronger broadcasters, this is certainly not a DX antenna.

Putting the SW1 on its own by the window with the telescopic whip extended produced similar if not better results in Europe than trying to use the radio inside the room with the cable trailing to the active antenna by the window. In North America and the Pacific where shortwave signal levels are low, we found the active antenna more useful, especially in an apartment. You could also try a piece of insulated long wire hanging out of the window, wrapping the other end round the telescopic whip of the radio.

Whatever you decide to do, it does not really matter...the active antenna is part of the complete package. Bearing in mind that on average the batteries last about 14 hours when used for around 3 hours continuous listening a day, you might be tempted to leave the active antenna and power supply at home. The radio on its own weighs just 232 grams and fits into a shirt pocket. The radio plus the plastic case, active antenna and power supply is half the size of a brief case and weighs 1620 grams.....seven times as much!

At a price of 799 Guilders in Holland, DM 699 in Germany, around 300 US dollars in North America, 39,000 Yen in Japan (Akihabara district of Tokyo) and £195 in Britain, the Sony SW1S is not cheap. You are paying for the small size packaging, and for the other accessories in the plastic presentation kit. In some countries the receiver is available as the SW1E package, which means you get a "travelers washing line antenna" instead of the active antenna and no power supply. Bearing in mind that the world-power supply with SW1S is so heavy, you might decide that the SW1E is a better value package. In Europe the SW1E package is retailing at around £135 in mid 1992. The fact that the set has no single-sideband reception or finer tuning on shortwave than 5 kHz may be no problem to you. But you should not expect this receiver to pick weak DX signals out of the background noise. Apart from the tuning steps, the set has excellent specifications for a shortwave portable receiver. If you want richer listenable audio and finer tuning you should consider other receivers in the same or even lower price classes.

SONY ICF-SW15

T his radio was launched in many markets in the second half of 1992. At press-time only a pre-production sample was available, so we have reserved our full technical report until the 1993 World Radio TV Handbook when we have purchased an off-the-shelf sample. However there are points which we want to raise at this time.

SIMILARITY TO THE ICF-SW20.

There is a remarkable similarity both in performance and looks to the Sony ICF-SW20. Although both sets are available at the moment, it looks like the ICF-SW15 will replace the ICF-SW20 in the long term. The price in Germany is DM150, around US$100. There's no longwave on this set, so if you rely on that band for listening to stations in Ireland, Britain, France or Germany, for instance, then you need to make another choice. The set has a dual conversion design on shortwave. It offers FM coverage, mediumwave (531-1602 kHz), and seven shortwave bands. On our sample we were able to tune in the following ranges: 5775-6430 kHz, 6880-7525, 9380-10030, 11615-12265, 15000-15650, 17500-18150, and 21325-21980 kHz. On the face of it, the coverage is better than the older ICF-SW20. There is far more out of band coverage, e.g. more frequencies outside the 19 meter band. However Sony hasn't done anything about the new 22 meter band (i.e. 13 MHz) so for the holidaymaker in Europe wanting to listen to some of the clearest and sometime the only audible frequencies from Austria, Holland, Russia, and Australia it is simply bad luck. Whilst the band spread on the radio is clearly better than on sets like the discontinued Sony WA5000, the exclusion of the 22 meter band is going to severely limit the success of this radio, at least in Europe. What a shame!

SONY ICF-SW20

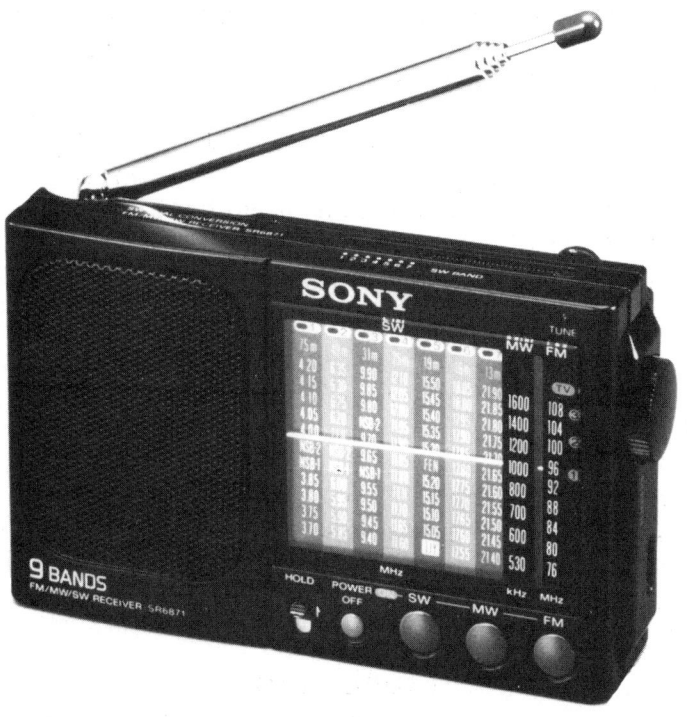

This radio has already been given quite heavy promotion in the trade press as an entry level receiver for the holidaymaker interested in being informed about news developments while relaxing on the beach. The receiver is about the size of a cigarette packet, indeed it's the same size as its more expensive brother, the SW1S. This radio though has no digital push-button tuning. It has a traditional tuning scale and silver pointer which allows you to read off the frequency to the nearest 25 kHz. Sony's new ICF-SW20 measures 11.5 by 7 by 2.5 centimeters, which will just about fit into a shirt pocket, depending of course what size shirts you wear! Later advertisements mention that it is about the size of an audio cassette. In any event, the set is extremely portable. It weighs a mere 214 grams, including the two penlight batteries which snap fit into the back of the receiver.

FACILITIES: DIFFERENT VERSIONS!

The ICF-SW20 is a broadcast-only receiver. There is no single-sideband to allow reception of amateur radio stations or ship-to-shore communications, but for the price being charged you can't expect such a facility. The set covers the standard mediumwave AM band, FM from 88 to 108 MHz, and 7 shortwave broadcast bands. These turn out to be the 49, 41, 31, 25, 19, 16, and 13 meter bands on the European version. The ICF-SW20 is designed to replace the older compact shortwave portables in the 4900 series. In doing so, Sony has introduced improved coverage of the frequencies either side of the official international broadcast bands. What is termed the 49 meter band actually runs from 5800 to 6400 kHz, or in the case of 16 meters from 17500 to 18100 kHz. That is the good news.

NO 22 METER BAND

The bad news is that in their attempt to keep the price low, Sony has omitted the 22 meter shortwave broadcast band. Back in 1979 the World Administrative Radio Conference decided that a new 13 MHz broadcast band would officially come into operation in the middle of 1989. In fact the final decision to open the 13 MHz band for business has been delayed indefinitely, because of problems at a more recent WARC. Nevertheless, under the motto "use it, or lose it", many countries are already using the 13 MHz band, and in general, the reception results are very good indeed. You can't pick up any 13 MHz broadcast band frequencies on the new Sony ICF-SW20, and we feel that is a serious omission, severely degrading the value of such a receiver. Some versions of this set appear to offer the 75 meter band instead of the 41 meter band. Although, in theory, the 41 meter band is not used for broadcasts to the ITU Region 2 (i.e. the Americas), in practice the band is very heavily utilized by broadcasters. The exclusion of the 41 meter band in favor of the 75 meter band is not a smart move.

CALIBRATION

For a receiver of this type, the tuning scale was quite accurate, though it can't compare to a digital readout display of course. The 4 centimeter speaker on the front of the receiver gives reasonable fidelity for news programs. There is a two position tone switch which allows you to cut some of the treble response if it sounds too shrill for your taste. There is an output for headphones, though because the FM section of the set is not stereo, the jack is a standard mono type. Unless you use an adapter, plugging a set of stereo headphones into the ICF-SW20 will only give you sound in the left ear.

Using the set in Europe and the Caribbean area we found it had good sensitivity. There is only one bandwidth filter in the set, but we found the selectivity to be fine for listening to the stronger international broadcast stations. Unlike some of the more expensive digitally tuned sets, you can sometimes reduce interference from interfering stations, by slightly off-tuning the desired station.

The dual conversion circuitry of the ICF-SW20 is straightforward. There is a small red tuning indicator lamp, but on shortwave this is more of a gimmick rather than a serious device to aid better tuning. In the European evenings when the 49 meter band is full of signals, the tuning light doesn't go off as you move up and down the dial. The set slightly overloads when the telescopic antenna is fully extended, though the problem is reduced to a minimum by simply partially collapsing the antenna.

CONCLUSIONS

To summarize: at a price in the United States of around US$90, and in Europe around £65 or DM200, the ICF-SW20 is a handy pocket shortwave receiver for the traveler. We think its use is marred by the inability of the set to tune the new 13 MHz broadcast band which will become increasingly important within the life-time of the set. You CANNOT make a cheaper shortwave portable by leaving out some of the broadcasting bands. With the possible exception of the 25 MHz or 11 meter band, the other 8 bands from 13 to 49 meters must be included.

SONY ICF-SW7600

Whilst Sony can justify its reputation for original thinking, the marketing managers in Japan haven't been very thoughtful when it comes to dreaming up a new type number. Sony still have models called the ICF-7600A, ICF-7000AW, ICF-7600D, and ICF-7600DS on sale in many parts of the world. These are completely different compact travel portables. There is also an ICF-7601 which sounds like a follow-up to a 7600 type radio. In fact it is an improved ICF-7600A but it is not better than the ICF-SW7600. Confused? Read the start of the review of the ICF-7601 for further information. But now let's look at the ICF-SW7600.

OPERATION

The ICF-SW7600 has a straightforward design and operation. To tune to 9895 kHz for instance, you press the AM button, the numbers 9 8 9 5, and then the AM button again. This executes the tuning request and the receiver tunes to that frequency. The keys on the keypad are well spaced out, so there is no chance of accidentally pressing two keys at once. The more expensive ICF-2001D has a small blip on the key for the figure 5. This is useful for visually handicapped users in helping them to orientate which key is which. There is no blip on the keypad of the ICF-SW7600.

The set jumps up and down the dial in 3 kHz steps between 150 - 531 kHz, 9 or 10 kHz steps between 531 - 1615 kHz (this is selectable by a small switch in the battery compartment) and 5 kHz between 1615 - 29995 kHz. The steps on FM are 100 kHz. Finer tuning on shortwave frequencies is desirable, and possible. A switch on the side can be set to activate a small thumb-wheel control for fine tuning +/- 7 kHz. Adjusting the thumb wheel has no affect on the frequency display. It is therefore a good idea to keep this thumb wheel in the middle of its track or the set is permanent off-channel by a few kHz. In the case of strong mediumwave stations this means it is possible to tap in 657 kHz but by turning the thumb wheel to one end the radio is actually monitoring the channel below, namely 648 kHz.

Sony has improved the Single-Sideband (SSB) reception on the ICF-SW7600 when compared to the ICF-7600DS. It now has a selector for upper or lower sideband. Adjustment of the SSB is via the fine-tuning thumb wheel. This is adequate for SSB transmissions you find on the amateur radio bands. But the control is not stable enough for radio teletype reception, or the technique of listening to one sideband of an ordinary broadcast signal (ECSS). But then this radio is not in the price range of the serious hobbyist and therefore should not be expected to offer these facilities.

The liquid crystal display is clear, showing the time while the radio is switched off and the frequency while switched on. The radio has a timer function, and can be set to operate a tape recorder. Note that the radio will switch a low control current not the mains supply. There is a output jack marked RECORD which has a very low level output. If you were using a "Walkman" type recorder, you'd need to connect the record out to the microphone input. Better results are obtained by connecting the headphone output to the line-input of the tape-recorder. This allows stereo recording on FM, although the speaker of the radio is then muted.

The set offers 10 memories which are programmed by pressing and holding the enter key, and then selecting one of the keys from 0 - 9.

PERFORMANCE

We made the following measurements on the sample:

Frequency Range	Sensitivity (in microvolts) for 20 dB S+N/N
2 - 6 MHz	2.1
12 - 17 MHz	10.6

Bandwidth	Dynamic Selectivity
+/- 5 kHz	26.0 dB
+/- 10 kHz	46.0 dB
+/- 15 kHz	51.0 dB

Blocking Level	4.2 mV

Image Rejection	-16 dB (at +/- 910 kHz)

These results show good shortwave performance for a radio of this type. We noted that an external long wire antenna is NOT recommended, especially in parts of the world with a high shortwave transmitter population (e.g. Europe). In more remote parts of the world a simple long wire antenna might be required for improved reception of higher frequencies. Any overloading when using the telescopic whip antenna can easily be reduced by collapsing the antenna somewhat.

The ICF-SW7600 is selective enough for international radio listening. However, the option of a second, narrower filter would have been handy. This is currently being offered on a similarly priced receiver from the Sangean company, the ATS-808. The Sony ICF-SW55 does offer two bandwidth choices, so you might also want to check the price differential between the ICF-SW55 and the ICF-SW7600.

FM performance is fine, and the addition of FM STEREO via the headphone socket is a definite plus. It seems strange though that the stereo signal is not available at the RECORD output.

A two position tone control is active on both AM and FM..it simply cuts the treble response when activated. The audio on the ICF-SW7600 is better than on the previous ICF-7600DS, offering better lower frequency reproduction and higher volume output. The DX/LOCAL switch gives about 18 dB of signal reduction when activated. We found it useful for mediumwave listening in Europe.

The radio has a cleverly designed ON/OFF switch which can be locked. This is useful when the set is packed in a suitcase and you don't want it to spring to life when knocked about. The radio works off 4 penlight batteries. A full set of batteries gave us 18 hours of good FM reception at reasonable listening volume. The batteries should be kept in the radio at all times, even when an external power supply is connected. This ensures that the stations programmed in the memory are retained. You have around 4 minutes to change the batteries before the radio resets itself.

CONCLUSION

Sony seems to have started to heed our advice and simplify the "7600" family when designating new type numbers. They realized that the improvements were being masked at the dealer level by people who can't easily spot the differences. The ICF-SW7600 is a well designed shortwave portable receiver ideally suited to international radio listening. It is a clear improvement over the ICF-7600DS, both in terms of better design and improved performance and features. The lower price (around US$220 in the US and £135 in Britain) makes it a strong competitor to other radios in the same price category.

SONY ICF-7601

Back in 1980, the Sony Corporation of Japan launched one of the first travel portable shortwave receivers designed specifically for international broadcast reception. Until that point, conventional designs had offered one or two shortwave ranges. But entire shortwave broadcast bands were compressed into a few mm on the dial. If you so much as breathed heavily, the radio danced onto the next station.

Then from Japan came the Sony ICF-7600, the model number given to this simple portable. It offered 5 shortwave bands, with considerable band spread. Whilst its size was convenient, it was expensive, overloaded very easily, and suffered from whistles despite its dual-conversion circuitry.

A few months after the ICF-7600 Sony announced its push button ICF-2001 receiver, in a much higher price bracket. Although the keyboard-tuned 2001 got most of the press attention, in terms of numbers sold the ICF-7600 was by far the winner. Two years later it was upgraded to the ICF-7600A. Battery consumption was reduced, and by upgrading the receiver design, overall performance was improved.

The biggest problem with the ICF-7600A was the restricted shortwave coverage. It did not take into account that many stations were operating just either side of the official shortwave broadcast bands registered at the International Telecommunication Union in Geneva. As a result, although it sold well, many people quickly became disappointed when the set refused to pick up the generally clear out-of-band shortwave channels.

Since then Sony has confused the market place in Europe, Asia, and the Middle East by starting a whole line of 7600 receivers, none of which have the same internal circuitry and specifications, and only seem related by their size, that of a paperback book. So the ICF-7600A is not the same as the ICF-7600D which, in turn, is not the same receiver as the ICF-7600DA.

Sony launched another "paperback book" sized portable radio onto the world market in 1990. It has the designation ICF-7601 and is priced at around US$120 in the United States, DM 225, or £89.95 in Europe. It has no key-pad tuning system, you simply find the stations using the traditional point and dial method. It is in fact the follow-up to the ICF-7600A. Sony has reacted to criticism and come up with a budget no-frills shortwave travel portable. It weighs 570 grams without the batteries, making it quite light.

The receiver does not, however, cover all frequencies between 100 and 30000 kHz. Instead, portions of the broadcasting bands are offered. mediumwave and VHF FM are standard. Then come the 9 broadcast bands, i.e. the 13, 16, 19, 22, 25, 31, 41, 49 and 60 meter bands. Finally on European models designated the ICF-7601L coverage of the longwave band is offered. On models sold elsewhere, a scale covering the broadcast bands of 75, 90, and 120 meter bands is provided. The 90 and 120 meter bands are reserved for stations in the tropical regions of the world.

TUNING

Tuning is quite straightforward. If you want 6020 kHz, you select the 49 meter band, and turn the tuning knob on the side until the pointer is in the region of 6020 kHz. The scale markings are to the nearest 5 kHz, though in practice, because there is slight backlash on the rotary tuning control, calibration is within 10 kHz. This is to be expected on a portable of this price range.

Unlike previous analogue models in the series, the ICF-7601 has ample coverage of the broadcast bands, including out of band channels. The only problems you might experience are finding Iran and Israel on their channels near 9 MHz, and if you buy the European version with longwave implying that you have to sacrifice the 75 meter band coverage. That is used by some European broadcasters such as the BBC and Swiss Radio International for continent-wide coverage in the winter. The lack of the 11 meter band is not really that serious as we head towards low sunspot years (94-97), and is offset by the fact that the new 22 meter broadcast band is tunable on this receiver. You should not really consider buying a shortwave receiver that cannot pick up 22 meters, or 13 MHz, for this is already widely used by many countries.

PERFORMANCE

The ICF-7601 offers a red light emitting diode which lights when a signal is being received, though on shortwave such a tuning aid is more of a pretty decoration than a serious help to tuning. Your ears are a far better tuning guide. One bandwidth filter is used in the receiver. It is just over 5 kHz measured at -6 dB, and offers a fair shape factor. The set copes better with overloading by strong shortwave signals than its predecessor, but using the set in Europe at night still leads to whistles generated as a result of some spurious mixing inside the receiver itself. Reducing the length of the telescopic whip may assist. A two-position tone control helps to reduce some of the familiar shortwave whistles when it is set to the news position. If you are someone who likes to listen to shortwave in subdued lighting, then the lack of a dial light might bother you.

Battery consumption of the four penlight cells is slightly above average. In Europe a 6 volt DC mains power supply is included in the price and works out far cheaper to use than alkaline batteries. Rechargeable batteries would be another option if you are forced to travel light. A lock switch is provided so that the receiver will not accidentally switch on while packed in a suitcase.

CONCLUSIONS

Overall this improved Sony ICF-7601 offers excellent value. Although it does not have digital readout, you can gently fine tune away some interference problems that are not possible on the far more expensive ICF-7600DA from the same company.

SONY ICF-2001D/ICF-2010

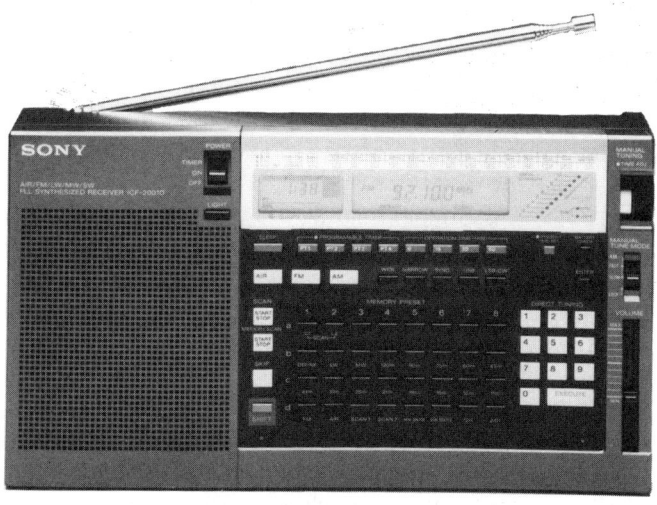

SONY ICF-2001D

The set bears two type numbers and was launched in early 1985. In North America it is known as the ICF-2010, in Japan and elsewhere it's called the ICF-2001D. That last type-number could have been confusing a few years back, because this Sony receiver has little resemblance to the ICF-2001, i.e. without the letter D after it. That radio was introduced in August 1980.

The model we tested was marked as the ICF-2001D. Put next to the old ICF-2001 it's very slightly smaller, measuring 29 by 16 by 5 centimeters, and weighing the same 1.8 kilos including the batteries. That's where the similarity ends, for the ICF-2001D certainly has a much larger array of push-buttons, 68 in all. At first sight, they might put off someone who's not all that technical at first. There are some familiar points, like a white calculator-style keypad, for directly keying in the frequency. If you wanted to listen to 11730 kHz, you simply punch in 1-1-7-3-0 and then push the "EXECUTE" button. If you can use a calculator, then this set is no problem!

MEMORY BUTTONS

Next to the familiar 11 keys though, are four rows, each of 8 buttons, representing the 32 channels you can store in the memory. These aren't marked 1, 2, 3, ...31, 32 as you might expect. It's more like reading a map, for the rows are marked A to D, the columns 1-8. A simple two finger operation stores any chosen frequency in the memory, which you're told is memory b8, or c5, depending on what you've selected.

GOOD COVERAGE

The ICF-2001D has wide coverage. It can be operated continuously between 150 kHz, right up to 29,999 kHz, that includes longwave, mediumwave, and shortwave. Plus there is the VHF FM band between 76 and 108 MHz, and, this is a new trend perhaps, the aeronautical band between 116 and 136 MHz. The air band is not available in some versions since it makes the receiver a forbidden fruit in many countries in Asia, where listening to the air band is considered an offense. But all that coverage, in a relatively small box, is impressive.

At this point, let's re-examine those 32 keys for the memories, in the four rows. Under most of the keys, is blue colored lettering, because like some computer and typewriter keyboards, each key has a double function. If you press and hold a key marked SHIFT, the blue colored functions become active, and for the most part it means you can get the set to jump immediately to the bottom of the longwave, mediumwave, the 120, 90, 75, 60, 49, 41, 31, 25, 22, 19, 16, 13, 11 meter shortwave bands, plus the FM and air band.

SCANNING FUNCTIONS

You can decide to scan between two chosen frequencies, stopping when a signal is picked up, or start looking through the 32 channel memory. In that case, the 2001D selects each memory with anything stored in it for 5 seconds, lets you hear it, and then moves to the next. If you like computers, and gadgets, this is a very flexible set.

You can also opt for manual tuning, with a conventional rotatable knob. It allows you to move in "steps" up and down the bands. You move in 50 kHz steps on VHF, 25 kHz on the air band, and either 1 kHz or 100 Hertz on short, medium and longwave, depending on what you select. Those steps are quite audible, and you may well find that since the minimum step is 100 Hz each time, that using this receiver for radio-teletype (RTTY) reception is rather difficult. It is less critical when listening to amateur radio operators on either upper or lower sideband, both possible on this set. Ideally we would have liked the tuning steps between 150-29999 kHz to have been 50 or even 25 kHz, though this would probably add to the cost.

The set will take account of the fact that the spacing between mediumwave stations is 10 kHz in North America, and only 9 kHz in Europe. This leads to an examination of the ICF-2001D's selectivity, or ability to pick out the station you want from the rest of the rabble. The wider the filter, (rather like a window looking out on the shortwave band), the better the audio quality, but the greater the risk other stations using frequencies nearby will also be heard. The ICF-2001D offers two settings, wide and narrow. We found the filters used to be fair for a receiver of this type. The "wide" setting is perhaps too wide, for in crowded bands you always seem to suffer from a 5 kHz whistle caused by adjacent stations. The "narrow" setting, around 3 kHz at -6 dB, is definitely needed, and still produces very acceptable audio when listening to news programs.

"SYNC" TUNING FACILITY

There's a button on this receiver which is the major high-point. It's marked "sync". Push it, and the receiver switches to either upper or lower sideband, but locks on to the station's carrier. It's an automated version of the old shortwave listener technique of switching to single-sideband and tuning very carefully to get rid of the resulting whistle. Why bother? Because the result is a signal less prone to the distortion due to fading, and it's very handy if there is a strong station on one side of the one you're trying to listen to. Press the "sync" button, and select either the lower or upper sideband, which ever gives clearer reception. The 2001D is rock steady enough for this idea to work well. The reason the feature is there is due to a chip developed for the AM stereo Kahn system. AM stereo has failed in the US, but the spin-off has been useful for shortwave listeners.

We noticed a design fault on the first examples produced in early 1985. The sync circuitry generated its own background noise. On lower frequencies, e.g. mediumwave or the 49 meter shortwave band, it's less of a problem. Later versions of the set (after 1986) had this problem solved. Just to make sure we tested another example of the set purchased in August 1992 and found the synthesizer noise had been considerably reduced. This enhanced the "sync" facility. On very weak signals, the sync facility fails to lock on to the carrier, but in 9 attempts out of 10, the system works well. The selectivity of the narrow filter was slightly better than on the previous model.

SENSITIVITY

Another important quality to check for is the set's sensitivity, or its ability to pick up weak stations. The results here are quite good. Our sample was very sensitive on all bands, and fortunately Sony have introduced an RF gain control. This means that if signals on the 49 meter band in the European evening get too strong causing distortion, you can reduce the signal level getting into the set. In addition there's a rather coarser switch marked DX-LOCAL, which we found was best left on the DX setting. On the old ICF-2001 you had to adjust an aerial tuning "thumb" wheel for best signal strength. This isn't needed on the ICF-2001D.

An extra use has also been thought out for the manual tuning knob. You can use it to set in the built-in 24 or 12 hour clock, and the versatile timer. That's clever, the only drawback being that there is no way to control an external tape recorder with the timer. That feature has been added on the new ICF-SW77.

Signal strength is displayed on a scale of 10 light emitting diodes, which also doubles as a battery indicator. The signal strength indication is nothing more than a guide, bearing little resemblance to the real "S" meter units. A light illuminates the frequency display for night-time use.

KEEP AWAY FROM STATIC

Seven years after its introduction there have been very few problems with the Sony ICF2010. But one of the most common faults is a sudden loss in sensitivity which renders the radio deaf. If you connect the radio to an external antenna, then you must disconnect this during any thunderstorm. Static build-up on the antenna can have a devastating effect on a field-effect transistor (FET) used in the front end of the radio. The FET in question is type number 2SK152. Sony service centers will replace the FET in question. If the radio is outside the warranty period you may want to try it yourself...that's providing you've done some electronics before. Remember that FET's don't like excessive heat so the soldering job has to be fast and precise. "Rambo" style repairs may end up costing more than having it done by a service engineer.

To get at the FET, remove the seven screws that hold on the back-cover. One screw is under the batteries and note that one screw is shorter than the rest..put it back where it came from. With the back off, and the speaker on your right, look for a separate printed circuit-board above and to the right of the speaker. This board lifts up, but take care not to damage the fine wires that go to the mediumwave ferrite antenna. On this board you'll find Q-303 which is the FET in question. Taking care the avoid static build-up (fully earthed soldering iron) replace the FET, taking care to get the leads the right way round.

The FET is available through electronic component outlets. Several WRTH readers note that SW Horizons, #61 - 52152 Range Road 210, Sherwood Park, Alberta, Canada T8G 1A5 has a stock of these FETs. Tel: +1 403 922 2872.

POWER CONSUMPTION

The old ICF-2001 back in 1980 quickly got a reputation for being very expensive to run on batteries. It took so much current the batteries got warm and lasted about 5 hours. The ICF-2001D takes three Size D batteries for the radio section, and 3 penlight cells for the computer section of the receiver. We measured current consumption as 150 mA at an average listening level, which is a vast improvement and quite economical. You're still better off using mains electricity if possible, and in many countries the set is supplied with a 4.5 DC power supply.

Looking at some of the other features, there's a headphone jack wired for stereo which is handy if you want to use the Walkman style headphones, and a tape-recorder output with a very low level. You have to use the "microphone" input instead of the "line" input on your tape recorder, which seems odd. A three position tone control is available, but the middle setting seems to be sufficient for most types of listening.

ANTENNA

The set has evidently been carefully matched to the built-in 120 cm telescopic antenna. There is provision for connecting an external shortwave and air band antenna, but we found a number of cases where connecting a 10 meter longwire antenna gave no improvement or even worse reception. Note the comments about static build-up above.....it could ruin the radio's sensitivity. The provision of a shoulder strap indicates it's intended as a portable receiver, and a power lock switch prevents the receiver going on accidentally in the luggage. The smaller, and cheaper, ICF-SW7600 is better suited for the traveler though, unless you plan to be in one place for a long time. The ICF-2001D has a considerably more audio power than the ICF-SW7600.

An active antenna, the Sony AN-1, is advertised as being suitable for the 2001 series. Under normal signal level conditions we did not find it made any improvement on shortwave signals. On mediumwave, the AN-1 was useless, as it is impossible to disconnect the ICF-2001D's internal ferrite rod antenna.

PRICES

In North America, the set is called the ICF-2010 to distinguish it from so-called "gray market" imports. In the United States we found the price was around US$360 in September 1992, whilst in Europe a price of around £289 is about average. In Japan (Akihabara district of Tokyo) we found the price to be 54,000 yen. Overall this set is excellent for the serious shortwave listener, and may have some applications for hobby listening too.

If you're interested in trying to DX using this receiver, you may want to consider filter modifications. Kiwa Electronics, 612 South 14th Avenue, Yakima, WA USA 98902 USA sell replacement IF filters specifically designed to improve the ICF-2001D. WRTH has not tested this modification, although we have received independent positive comments from readers.

SONY ICF-SW77

The Sony ICF-SW77 was first shown at the Friedrichshaven amateur radio fair in June of 1991. That was a hand-made sample, and production versions of the set appeared in November of 1991. Sony Germany was the first Sony outlet to subsequently withdraw the radio after just four weeks on the market. Sony dealers in Holland, US and Canada followed suit. However Sony in Britain did nothing and even today they have advised several WRTH readers that problems never existed or that they didn't affect radios sold in Britain...they did. In February 1992 the ICF-SW77 was relaunched, but since it looks identical how can you tell it apart?

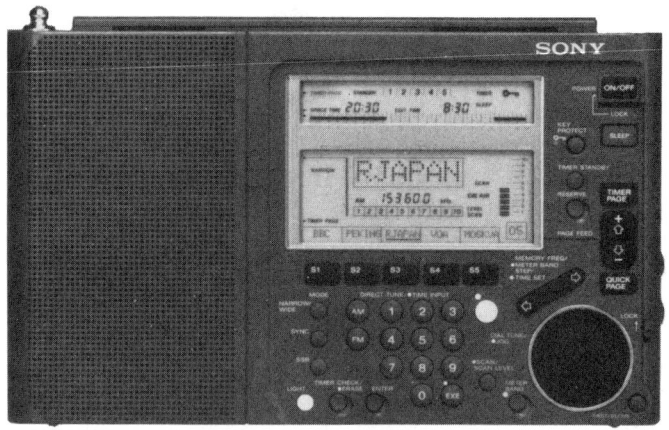

WHAT WAS WRONG?

When we tested the ICF-SW77 in late 1991 we found two major problems with the receiver. The radio's synthensizer was extremely noisy, leading to very poor performance on longwave. Side by side against the ICF-2001D, the older set could receive longwave station BBC Radio 4 perfectly, whereas it was difficult to follow the conversation on the ICF-SW77. The second problem we encountered was that the synchronous detector didn't perform well. When the sync button was pressed the audio response in both the upper and lower sideband was severely compromised. The bass and mid-range response in particular was cut drastically which did not lead to an improvement in intelligibility.

In our regular survey of dealers we asked Richard Robinson, manager Electronic Equipment Bank in Vienna Virginia USA for his findings. He had an extensive stock of the first batch of radios. He reported that around 5% of the sets had a software problem which led to the memories becoming full before they should have. When around 120 stations had been stored, the radio would not accept any more channels.

NEW FOR OLD: THE BLACK DOT

Once the relaunch had taken place we waited until May 1992 before buying another sample at a shop in Utrecht, Holland. This radio looked identical to all the previous versions, but performance wise it was very different. The synthesizer noise has been reduced considerably, and the problem with the synchronous detector cutting the audio response had been fixed. We understand that in Europe at least the new improved Sony ICF-SW77 were marked with a special black dot. This dot doesn't appear on the radio, but on the serial number sticker attached to the

cardboard box. Our example is illustrated below.

The following results are based on measurements of the new "improved" ICF-SW77 purchased in May 1992. When compared to the older ICF-2001D, you can summarise this new radio as follows.

Good points
- Improved ease of tuning.
- Stereo on FM
- Pre-programmed stations.
- Slightly better audio than 2001D
- Single set of batteries
- "Sync" control is easier to operate.
- Pre-programmed frequency information can be updated.

Bad points
- Synthesizer noise slightly higher than ICF-2001D. Reception of distant stations on longwave and mediumwave is more difficult.
- Background noise of the amplifier is too high. This is annoying when listening to FM on stereo headphones.
- No air band coverage 116-136 MHz in contrast to the ICF-2001D.
- Some pre-programmed frequency choices are not correct. Italian channels are wrong. VOA doesn't use some of the channels given.
- Instruction manual will be very difficult to follow for first-time users. Descriptions of the Quick-Page and the programming of the timer need much more explanation. Some advice missing: e.g. the find out the current UTC time, just listen to any international broadcaster on the full hour. Time on all international broadcasting stations is given in UTC/GMT.

DESCRIPTION

The new ICF-SW77 takes up the same volume as the 2001D. It has rounded sides though, and it is easier to hold thanks to better styling. It uses 4 size C cells, other wise known as the R14 size, to power it. So it's good-bye to the 3 large Size D cells that ran the 2001D, and the three penlight cells that were needed to run and back-up the computer section of the radio. That's good news because when the set was transported the large batteries would bang against the penlight cells, and after a while many owners reported problems getting the back-up batteries to stay in properly. The batteries are smaller so they don't last as long. The 2001D uses 130 mA at average listening volume, the 77 uses about 20 mA more from the power source than the 2001D, and 150 mA means that the batteries will last for a shorter period. In our tests a full set of batteries went for 14 hours using AM at a comfortable volume. You're better off using the AC power supply if possible.

POWER SUPPLY

Sony, in common with many other Japanese companies has now introduced a new DC power socket on its radios. The center pin is now positive, and the size and color of the socket is dependent on the voltage the radio needs. The SW77 works on 6 volts DC, and has a yellow socket to indicate that. If you have an old power supply, which has a plug with the + and - the other way round, don't worry. The plug on your old power supply won't fit the radio anyway...so no chance on any problems.

The 2001D was often criticized for its rather tinny audio, especially when listening to local stations on FM. Sound is extremely subjective of course, but most users will agree that the audio on the SW77 is richer and fuller than the 2001D. The speaker is bigger and now oval shaped, and there are separate bass and treble controls. On both AM and FM you can get a rich powerful sound.

HISSY AMPLIFIER

If you want good stereo from local FM stations that can be done on the SW77 if you connect a pair of good headphones. However, on our example, the amplifier hiss was annoying on a good pair of stereo headphones. The level was MUCH higher than the ICF-2001D.

TUNING

The 2001D had a front panel which included a matrix of 32 push-buttons. Daunting at first, these turned out to be memory buttons. You could store the frequency and mode of a favorite station at the touch of two buttons. But you had to make a note on a piece of paper somewhere as the name of the station under button A7 for example. Clearly the concept of the SW77 was to get rid of all these mechanical memory buttons and create an electronic note-pad which could store the text of a station name, its frequency and mode. They then put this electronic notepad inside the receiver, and provided forward and backwards buttons allowing you to page through the notepad as it were.

The set is being sold in around 6 versions, depending on local restrictions. The full coverage version offers continuous AM coverage from 150 kHz up to 30 MHz, plus FM from 76 to 108 MHz. The set can receive signals in AM, USB, LSB, and the synchronous detection feature of the 2001D is also on the SW77. However, in some countries, such as Saudi Arabia, SSB coverage and a portion of frequencies between 285 and 531 kHz and above 26100 kHz is blocked out. Sets sold in Italy have some restrictions built in to them too. What with 1993 round the corner, it makes you wonder how long such restrictions can remain in force. But still, Sony's not to blame for this.

JOG SHUTTLE

Not surprisingly, you can type in a frequency directly. There's a manual tuning knob, now re christened a "jog shuttle" control, which allows you to move up and down the AM dial in 50 Hz. Resolution on the frequency display is 100 Hz; you still get that "chuffing" sound as it does so.

It is important that you set the internal clock to the correct local time, and tell it the difference between your local time and UTC. That's because the radio comes with schedules of some 26 international broadcasters already programmed into it. By that I mean, not only the frequencies being used, but the time they are scheduled to broadcast on those channels. Sony tell us that radios shipped to Europe have a different set of channels programmed into them than radios shipped to Asia or North America. That's logical. On our European version we conclude that about 80% is correct. You push a button marked BBC or VOA, and most of the time that's what you'll hear. Larger broadcasters with lots of transmitters are able to create so-called in-house channels. The BBC makes extensive use of 9410 or 15070 kHz for instance.

The European frequencies for stations like Sweden or Denmark are also programmed, although you're not going to hear them outside the continent. Luckily you can not only add your own favorite stations to the electronic logbook, you can also change the pre-programmed channels. You use the manual tuning knob to select letters of the alphabet. The set also has an ingenious timer, more versatile than the 2001D, and there is provision via an optional extra for the radio to control a tape-recorder.

ONLY FAIR PERFORMANCE

We put an SW77 and 2001D side by side. On our example at least, the synthesizer noise from the new receiver was noticeably higher than the old one... especially on medium and longwave. Switching between upper and lower sideband in the sync mode is more convenient on the ICF-SW77. Both receivers offer a narrow AM mode which you'll need if you're listening to a station and there's heavy interference from a transmitter 5 kHz away.

CONCLUSIONS

If you're expecting major improvements to reception from this new set, the conclusion is that there's not much difference. The SW77 has slightly better audio quality though, finer tuning, and a whole variety of extra tuning aids which you may or may not find useful. 162 memories gives you an enormous database of listening suggestions and updating is easy once you wade through the instruction manual. The push buttons on the receiver have a clear profile above the front panel which will be handy for visually handicapped users.

In short, this is a nice set. You're certainly getting a lot of technology for the price and Sony have gone out of their way to improving tuning convenience. The price is higher than the ICF-2001D so you have to decide whether the extra tuning facilities are worth the extra money. In the United States we found the price was around US$490 in September 1992, whilst in Europe a price of around £349 is about average. In Japan (Akihabara district of Tokyo) we found the price to be 70,000 yen.

SONY ICF-SW55

There's a danger that the ICF-SW55 will be dismissed as a budget version of the SW77. Whilst the ICF-SW55 doesn't have synchronous detection, our WRTH tests show that it offers overall better performance than the ICF-SW77. To someone new to international radio, the front panel of the ICF-SW55 looks like a luxury electronic game. The front panel is dominated by a sophisticated 95 x 55 mm liquid crystal display. The data which appears includes a world map, UTC & local time, station name (when programmed), frequency, mode, and signal strength. But if you're prepared to read the instructions, the radio quickly becomes extremely easy to use. The extra facilities offered, compared with the more simple ICF-SW7600, allow you to get far more out of shortwave listening for a reasonably small extra investment.

CLEVER DESIGN

We gave the radio to several non-technical people to experiment with. Several commented on the clever compact design. The ICF-SW55 is only slightly larger than the ICF-SW7600 (for the record it measures 125 x 195 x 35 mm), but the designers have been able to spread out the controls by using a unique loudspeaker construction. The speaker is actually mounted at the back of the set, and a curved duct is used to guide the sound through a "letterbox" on the front of the radio. Openings at the back are important for the bass response, so the radio sounds best when propped up at an angle on a table-top. This is easy using the built-in stand. The radio sounds no different (or worse) than sets of a similar size using a conventional front mounted loudspeaker.

TUNING

The software used for the direct access tuning is more user friendly than cheaper Sony sets. On the 7600 series, for instance, pressing key "1" called up Preset One from the memory. To punch in a shortwave frequency you first selected AM, then tapped in the frequency, and then pressed AM again. On the ICF-SW55 you simply type in the frequency and press the execute button. You only need to press the AM and FM buttons when you switch from FM reception to long, medium or short waves, or vice-versa. The receiver remembers the last AM (or FM) frequency selected, which is useful when checking for a program running in parallel. The receiver also decides a default mode, depending on the frequency selected. Tap in 9895 kHz and the set selects AM wide for broadcast reception. Tap in 10145 and the set selects upper sideband as the default. These mode changes also occur when manual tuning is selected.

Manual tuning is achieved with a "jog shuttle" rotary tuning knob. In the so-called "FAST" position, manual tuning moves in 1 kHz steps on AM, 500 kHz steps on FM. If you select the "SLOW" position the steps reduce to 100 Hz steps on AM, 50 kHz on FM. Whilst even the fine steps are audible as a slight "chuffing" sound, they are small enough to allow easy tuning of single-sideband signals (in contrast to the ICF-7600), although the steps are occasionally a little too coarse for easy deciphering of RTTY signals. The manual tuning can be locked, whilst allowing direct keyboard access to remain active.

Two buttons allow you to jump up or down the shortwave dial in 5 kHz steps, 9 or 10 kHz steps on mediumwave, 3 kHz on longwave, and 50 kHz on FM.

NAME THAT STATION!

Sony's advertisements highlight what they call "Station Name Tuning", also available on the ICF-SW77. Research shows that people remember names of stations much better than frequencies. Hence a radio which displays the name of a station, as well as a number of alternative frequencies for that broadcaster, would seem to be a logical step. Grundig tried it successfully on their Satellit 500, and have extended the idea on the new Satellit 700.

The Sony ICF-SW55 gives you 25 electronic "pages". Each page allows you to define one name, and up to five (frequencies + bandwidth + mode), making a total of 125 channels. Start by calling up a blank page. Letters can be entered into the receiver by pressing the orange "Label Edit" button.

The buttons of the numeric keypad then represent letters of the alphabet. Since there are 26 letters of the alphabet, plus some extra signs such as *, you may have to push each numeric key up to three times to get the letter you want. Numbers can also form part of the name, e.g. "BBC R.1". It doesn't take long to program the names you want.

Since there is one 8-character name per page, you can either decide to store the name of a station, plus up to five favorite frequencies. Alternatively, you could tap in the name of the country, and store the frequencies of five favorite networks. You could also store one station, one frequency, but with options for AM wide, AM narrow, LSB, and USB under the memory keys. Paging through the system is made easier by a "Last Page" button which lets you spring back to the page you were programming.

A separate "page" is used to program the versatile timer. You can set up to five stations into the timer, each with different frequencies and/or modes. You can select the on time (either local time or UTC) to the nearest minute. The off-time is anything up to 199 minutes later. If you want, you could program the set to play the first two minutes of station A (with say news headlines) and switch over to another station for a local weather summary. A separate alarm function can also be selected in the same way. A socket on the side of the ICF-SW55 can be used to activate a suitably adapted tape recorder.

The radio comes complete with some international broadcasters and their most popular used frequencies already programmed into the memory pages. BBC, VOA, Radio Moscow, Deutsche Welle, Radio Netherlands, and Radio Veritas were just some of the stations. The stations stored by the factory will depend on whether you buy the set in Asia, Europe, or the Americans. For the most part, the choice of channels by Sony seemed logical. You can change any of the factory set frequencies very easily, although if you press the RESET button in the battery compartment the set reverts to the factory settings.

POWER

To get the set working you need to put in four penlight batteries. These batteries are also used for backing up the memory and the dual-time clock. You have about three minutes to change the batteries before the clock resets and the memory is erased. A button allows you to check the condition of the battery, and a symbol in the display gives advance warning of low power. The set consumes around 80 mA at a comfortable listening volume. In our tests, a fresh set of alkaline batteries lasted a total of 9 hours with listening sessions of around 90 minutes at a time. Batteries are some 200 times more expensive than household current. A dual voltage AC power adapter is included with the set in most countries. Note that this receiver (like the SW77) uses the new Japanese standard DC socket, which is smaller than previously. The center pin is positive, not negative as with many power supplies.

A small world map is shown in the Liquid Crystal Display. Two black lines show the local time zone selected. If you set the clock correctly to your local time, then it's possible to check the current time in several capital cities around the world. Of course the radio can't compensate for the fact that summer and winter times start on different dates around the globe. A button also allows you to display areas of the world currently in darkness. Another button compensates for local summer or daylight time.

The radio has a 20 segment signal strength meter, although this is just a rough guide rather than a calibrated instrument. While tuning up parts of the shortwave dial, the meter rarely leaves the maximum signal strength position.

SELECTIVITY - NO SURPRISES

The ICF-SW55 has more than adequate sensitivity on all bands. The synthesizer noise is significantly lower than the SW77, this being especially noticeable on longwave. AM selectivity is switchable between WIDE (6 kHz at -6 dB) and NARROW (2.8 kHz at - 6 dB). Nothing new about the standard filters installed. Selectivity is good for a portable.

Connecting an external antenna is possible, and this disconnects the internal ferrite rod and telescopic whip. In Europe an external antenna is not really needed unless it can provide signals which are less prone to man-made interference. If used, best results are often obtained with the sensitivity switch set to LOCAL. The dynamic range of the set is good for a portable, but it can't cope with the huge signals generated by many (active) antennas. In low shortwave signal strength conditions (e.g. in North America or Asia), a short length of wire (or the supplied "clothes line" antenna) may boost the strength of higher frequencies.

A pair of stereo headphones can be connected to the side of the SW55. You can switch between stereo and mono on FM, although there is no indication of stereo on the liquid crystal display. A three-position switch serves as a tone control, cutting varying degrees of treble response in two of the settings.

CONCLUSION

In short, the ICF-SW55 is an excellent portable receiver with a lot of built-in features. It calls into question what you really gain by paying significantly more for a ICF-SW77. If you're interested in either of these new sets then take the time to really compare the difference. It offers significantly more features than the similarly priced ICF-SW1, except that it's not the size of a cigarette packet. Price in the US is around US$370 as of October 1992. In Japan (Akihabara district of Tokyo) we found the price to be 44,800 yen. The price in Europe seems to be around £250.

SONY CRF-V21

This radio was first launched at the 1987 ITU Telecom meeting in Geneva. After several modifications to the original circuitry design, thereby improving the receiver performance, the radio is available world-wide. It is one of the most difficult radios to review because it offers so many features, and it would take a good four to five months to get to know all the functions by heart. Although Sony have worked hard to come up with a simple instruction book, most features need quite a bit of explanation.

Let us divide the discussion into three main sections:
    a) AM/FM receiver performance
    b) Weather satellite reception
    c) Radio Teletype decoding.

Bearing in mind the price tag, if only one of these features interests you, then there are much simpler and better solutions, albeit not so compact.

AM/FM PERFORMANCE

The AM section of the radio runs between 9 kHz and 30 MHz continuously. Unlike other receivers in this price range, the coverage then stops, resuming at 87.5 MHz (76 MHz outside Europe) up to 108 MHz for listening to FM broadcasters.

Simple tuning is achieved by selecting the mode and then tapping in the frequency on the keypad. The software here is user friendly, and clearly based on Sony's successful ICF-2001D (or ICF-2010). We measured the following selectivity figures

Selectivity:

Mode	AM Wide	AM Narrow	SSB	FAX/RTTY	NBFM
-6dB	3.0	1.45	1.4	1.4	7.0
-60dB	8.5	3.9	3.1	3.1	14.5

As far as sensitivity is concerned, the radio really does cover the very low frequencies, the time signal station on 77.5 kHz coming in well in Holland where the receiver was tested. The radio is deliberately less sensitive on mediumwave to avoid overloading from local stations. shortwave sensitivity was good, and FM sensitivity was excellent (FM reception is only in mono though, even on headphones). shortwave selectivity was fair. On AM you have two bandwidth choices, only one in the RTTY and SSB modes. AM narrow has a reasonable shape factor, but the resultant audio is very muffled.

Like the 2001D, the CRF-V21 offers true synchronous detection, a small LED showing whether the signal is locked in. We found this feature greatly reduces the effects of fading on a signal, and we found the improved sync design in the CRF-V21 was able to stay locked on weaker signals than is the case with the ICF-2001D. On strong broadcast signals, the audio is better than the 2001D, especially if an external speaker is connected. However, unlike most professional communications receivers, you have no facilities for pass-band tuning or notching out heterodyne whistles. This is a serious disadvantage when chasing weaker signals, putting the receiver's overall performance on line with radios costing around US$1200.

The receiver comes complete with two antennas for AM/FM reception. We would describe these as consumer quality rather than of a professional communications variety. The active antenna does not overload the input circuits of the CRF-V21 simply because its output is very low. On mediumwave the performance is only fair. In Holland we could log all the local mediumwave stations on twice their operating frequencies, these being internally generated by the active antenna. Much better overall results were obtained using a 10 meter longwire antenna or a T2FD.

TUNING

The radio has a large manual tuning knob which shifts the frequency by 10 Hz or 1000 Hz steps (switchable) below 30 MHz, and 25 kHz steps on FM. The LCD frequency display only has a resolution down to 100 Hz. The steps are small enough for easy SSB and RTTY tuning. Two buttons underneath the tuning knob allow you to jump in greater steps, 1, 3, 5, 9, 10, 12.5, 25, or 50 kHz at a time. The tuning step is always displayed on the screen for easy reference.

Most of the receiver's current status is shown on the LCD screen which is a focal point of the radio. However this screen is very poorly lit. The plastic which covers the screen reflects daylight badly, so if you operate the radio in subdued light with light falling on the radio from behind the operator (or from the side) part or all of the display is difficult to read (especially fine detail on the spectrum analyzer). The manual offers a solution in the form a reading light which is normally parked at the top of the radio, and which is activated by swinging it down in front of the screen and pressing the LIGHT button. However, the best angle for the lamp corresponds with the position where it obscures the view of the screen, and if you do not tune the radio for a period of 30 seconds, the light goes out. This is a serious design flaw, made worse by the fact that it is easy to mistake the reading light for a carrying handle. The CRF-V21 weighs 9.6 kilos with the rechargeable battery installed, and the light will therefore bend or even snap off if you lift the receiver with it. The manual warns of this fact, pointing out the built-in carrying handle at the top of the set.

Signal strength is shown on the main display. It is possible to set the squelch level by moving a pointer on this display. If this is set too high then the set mutes altogether. This is clearly explained in the handbook. The signal strength readings are clearly only a guideline, falling well outside recognized norms for an "S" meter. Whereas S8 should correspond to a signal level of 25 microvolts, and S9 by an input of 50, we noted the Sony gave a reading of S8 at 18 microvolts, and was well above S9 for an input level of 38 microvolts.

THE SPECTRUM ANALYZER

The LCD display can be switched to show a part of the radio dial, i.e. working as a spectrum analyzer. This receiver was the first consumer set to offer such a facility (back in 87 when it was first shown) but ICOM now have the facility on the R9000 too. In Sony's case however, the width of the "window" is either 200, 1000, or 5000 MHz. Although resolution of 1 kHz is possible on shortwave AM, we find that to be rather coarse in comparison to Icom's solution, as well as having a much slower response time. You cannot really detect activity in the sidebands that is possible with the more expensive R9000. You can print out the results onto paper for future reference. The analyzer has a reasonable dynamic range, detecting signals as low as 5 microvolts (using the low impedance antenna input). The radio mutes while carrying out a scan.

POWER SUPPLY

This receiver works from the mains power (via an ACP-88 adapter) or from a rechargeable Ni-Cd battery pack (as used in Sony's Video 8 cameras). The power pack is more than a little black box you plug in the wall. It is huge, and weighs 2.8 kilos! It certainly supplies the radio with enough power without getting warm, but it is clumsy as a result. The low-voltage end of the power unit is in the form of the NP-22 Ni-Cd battery pack, so you can either plug this into the side of the V-21, or use it to charge a battery which takes about 2 hours...you cannot do both at the same time.

FAX RECEPTION

The thermal printer in the CRF-V21 has excellent resolution, capable of definition that is difficult to reproduce accurately in a book like this. The set can either print out weather maps picked up on longwave (which have a lot of weather information marked on them) or by connecting an optional satellite dish and down converter, either the METEOSAT, GOES, or GMS weather satellites can be successfully decoded. The receiver we tested erroneously showed GOES in the display instead of METEOSAT, but printed out the pictures nevertheless. The satellite frequencies are pre-programmed, so should these change in the future there could be problems. Likewise, such pictures are not scrambled at present, but there is no guarantee that this will continue throughout the nineties. The satellite reception was easy to set up for even a non technical person, and the instruction book is very clear in this respect. The satellite signals are received in the 1690 MHz range, so it is essential that the satellite dish remains stable, pointing to the geostationary signal source. Both FAX and RTTY data can be stored for later playback by recording the data onto a standard cassette tape-recorder (not built in!). This system works well, but it is slow if you are used to storing data on a floppy disk.

RTTY RECEPTION

The radio-teletype capabilities of this set are aimed at the commercial user, rather than the amateur. Baudot code at 45, 50, 57 & 75 Baud can be decoded, as well as ASCII data running at 110, 200, 300, or 600 Bits per second. But do not expect the unit to read Morse code (even commercial CW) or popular error-correcting systems such as AMTOR (as used on the amateur radio bands). Some of the RTTY parameters are automatically detected by the V-21, others can be manually adjusted (e.g. the polarity). We found the unit was able to make perfect copy even when the signal went through a deep fade or got very noisy, indicating good error correction software.

RS232C CONTROL

Most versions of the radio have the option for computer control. PC software has not been developed in Japan, but Sony's headquarters in The Netherlands has put an MS-DOS control program together. The control codes are well described in the manual, so if you have some computer experience, writing your own program is not too difficult.

CONCLUSION

This radio is versatile, offering a wealth of functions in one case. Compromises have to be made, and if you only want one of the functions offered (e.g. shortwave reception or RTTY decoding) there are cheaper solutions. The thermal printer is excellent, but note there is no way to display FAX data on a screen. There are systems designed for home weather reception that will store METEOSAT pictures and play them back on a monitor in sequence. Cloud formation and movement is easier to follow in this way than looking at black and white maps placed next to each other.

At around £2700 or DM10,000 in Europe, or US$4700 in the United States, this radio is expensive, maybe too much for even the most dedicated enthusiast. It has already found a market in some industrial sectors (e.g. farmers wanting to know weather details).

YAESU FRG-8800

Yaesu's popularity in the shortwave broadcast receiver market has risen and fallen over the last 15 years. In the late 1970's the FRG-7 was one of the most popular entry level shortwave receivers around. In the end Yaesu said they stopped making them because the die to punch out the cabinet wore out and they realized it was time to launch something new. The FRG-7000 was quickly followed by the FRG-7700, both of which are now obsolete. The FRG-8800 arrived in 1985 amid quite a launch. It was the first to make considerable use of microprocessor techniques, although static discharge from the antenna or from the user can sometimes cause the radio to "freeze". The remedy involves resetting the receiver. The FRG-8800 is now to be phased out with the arrival in late 1992 of the FRG-100. However, since the FRG-8800 is still in many catalogues we have included test results in this edition.

SENSITIVITY

High sensitivity figures on shortwave aren't as important as the manufacturers claim. After all, the atmospheric noise from a decent outdoor antenna can run into the microvolts. We made the following measurements:

Sensitivity 10 dB s+n/n for FRG-8800 (all in microvolts)

Frequency (MHz)	AM Narrow	SSB/CW	FM
0.2	4	1	1.6
1	3	0.7	1.0
5	0.6	0.15	0.19
10	0.6	0.15	0.19
20	0.8	0.20	0.27
29.9	0.8	0.20	0.25

The figure of 10 dB signal+noise, divided by noise is only just intelligible. For usable speech we usually need a ratio of about 20 dB. So we tuned the radio to 10 MHz and measured how much signal was needed to get certain s+n/n ratios. This gives the following:

Signal Strength @ 50Ω in microvolts	S+N/N ratio (dB)
0.06	3
0.1	6
0.2	13
0.5	19
1	21
3	26
5	30
10	34
100	36
1000	38

S METER

The FRG-8800 has a liquid crystal display which replaces an old-fashioned style needle. Other manufacturers later followed Yaesu's lead. We found that the meter wasn't too bad for the price. Under S3 the meter had a tendency to show signal strengths below what they really were, and above S6 the meter was too generous. But we have seen a lot worse.

SQUELCH

The FRG-8800 has a squelch which is designed to silence the radio when no signal is received. This is ideal for narrow-band FM reception, but the squelch also works in other modes. The range of the squelch is important so that you can set it to let through weak stations or just the strongest. In the SSB mode we made the following measurements at 5 MHz.

Squelch	Cuts In	Cuts Out
Lower Limit	0.6μV	0.4μV
Upper Limit	750μV	620μV
With RF gain at minimum	300 mV	250 mV

RF GAIN CONTROL

The FRG-8800 is fortunately fitted with a continuously variable RF gain control. That is useful bearing in mind the huge powers used by shortwave broadcasts. At this point, if you haven't already read the introductory chapters about receiver specifications, you might want to stop here and do so. They will explain why the use of the RF gain control is useful. Try turning it back a bit and turning up the volume control. You may notice that interference levels drop and the signal is much easier to understand.

RF Gain control FRG-8800

Gain Position	0	Half	Three-Quarters	Full
Sensitivity 10 dB S/N (μV)	0.15	0.36	3	60
Attenuation (times)	0	2	20	400
Attenuation in dB	0	6	10	52

STABILITY

It may seem a bit strange in the 90's, but receiver stability was a common problem in the 70's, especially those with valves in them. The fore-runner of the 8800, the FRG-7700 was very sensitive to temperature variations and that was often an annoying problem when using the set for radio-teletype reception. The FRG-8800 is much more stable. After an initial warm-up period of 30 minutes, the set drifted less than 120 Hz. That's a good figure, especially as the frequency chosen was 29 MHz, at the high end of the receiver's coverage. The fine tune control allows you to shift frequencies continuously between +550 Hz and -690 Hz and therefore has a range of 1240 Hz.

AUTOMATIC GAIN CONTROL

The AGC on the FRG-8800 turns out to be excellent so that signal variations between 0.5 microvolt through to 100 millivolt are kept at a constant volume. The AGC constant is switchable between two settings.

BIRDIES

Every receiver produces some internally generated signals which show up as unmodulated signals on the dial. Weak whistles will disappear under the noise floor of the receiver, but if strong birdies appear then these will be noticeable and may block out weak signals. We made a series of measurements in a Faraday cage to determine birdies on the FRG-8800. In total we measured 40, which was rather higher than expected. Some of the frequencies of birdies stronger than 1 μVolt include: 4249, 5312, 11250, 11308, 14925, 18000, 22616, 22957, 23985, 26000 and 29982 kHz.

SELECTIVITY

The two bandwidth filters used in this receiver are on the wide side. The skirt selectivity is not bad, but the - 6dB points are at 3.9 and 8.4 kHz (NAR-

ROW and WIDE respectively). That is too wide in both cases. It leads to some excellent audio on strong signals but limits the receiver's use for serious DXing. That move came as a surprise because the FRG-7700 came with three bandwidth options.

EXTRAS

The radio offers a built-in clock, but the readout is shared with the frequency display and it doesn't show seconds. The timer functions are limited although Yaesu does offer a computer aided tuning software package to assist in this respect. The radio comes complete with a well-written manual, one of the best and most comprehensive for a long time.

DYNAMIC RANGE

The FRG-8800 was tested following the CEPT standard. It came out of the test with a dynamic range of 90 dB, compared with 80 dB for the Kenwood R-2000 and the 96 dB we measured for the Icom IC-R71. This is a good figure, although not in the same league as the Kenwood R-5000 or NRD-535.

CONCLUSIONS

Yaesu is always one of the top three manufacturers in the amateur radio market producing some of the best transceivers money can buy. They clearly regard the shortwave broadcast market as important too, although there is only one receiver in the marketplace at any one time. The radio is currently priced at US$700 in the United States. In Japan (Akihabara district of Tokyo) we found the price to be 93,200 yen and around £639 in Europe. However, this is expected to change in late 1992 as the new FRG-100 is brought into the market. The FRG-8800 remains an excellent receiver for the shortwave listener, even in later years when it appears on the second-hand market.

YAESU FRG-100

Yaesu's newest receiver is designed to compete with the Drake R8 and Lowe HF250. It is based on the receiver section of the new FT890 transceiver and should offer better dynamic range than the existing FRG-8800. Launched at the JAIA Ham Fair in Tokyo on August 20th 1992, we hope to have a production model to test in time for the 1993 World Radio TV Handbook. More information from Yaesu Musen Co.Ltd, C.P.O. Box 1500, Tokyo Japan. Yaesu USA 17210 Edwards Road, Cerritos, California 90701 USA. Yaesu Europe, Snipweg 3, NL-1118 AA Schiphol, The Netherlands. Tel: +31 20 6037299. Fax: +31 20 6480445.

NOT IN THE BUYERS GUIDE: TESTED & REJECTED

We examined and rejected the following models from this Receiver Survey because the short-wave coverage was obviously an extra. Performance between 3 & 30 MHz was substandard: AudioSonic TKS-326, TK-333S, TKS-342, TK-344F, TKS-350, TKS-354, Grundig Music Boy 170, Grundig Prima Boy 70, Grundig Boy 40, Grundig Concert Boy 230, Grundig Cosmopolit, Panasonic GX80, GX50, GX30. Philips D2615, Philips D2225, Philips D8184, Philips D8478, Sony AIR-7, Sony WA-6000, Supertech WE-9, Supertech WE-110A.

We also rejected all models with the brand name "Marc", "Magton" & "Tokyo Skylark", "Yoko" & "Frontech".

CHAPTER 9
SHORTWAVE RADIO ON THE MOVE
BY DAVID ROSENTHAL

Around ten years ago in Europe you could buy a shortwave converter from either the companies of Grundig or Becker. The converters fitted between the car antenna and a standard car radio with a medium wave AM band. By pressing a button on the converter box reception of medium wave was suddenly transformed into shortwave reception of, e.g. the 31 meter band. These converters have long since been discontinued, and you would be hard pushed to find one on the second-hand market.

The converter device did a reasonable job, but tuning the dial was more by feel than anything else. You had to know by heart that a particular station was down the bottom end of the 49 meter band, and little bit to the right was another station, and so on. This is not much use to a novice. In addition, the placing of the car radio antenna was crucial. If it was put too near the engine, the radiation from either the mechanical or electronic ignition system drowned out all but the stronger shortwave broadcast signals.

The shortwave performance of one car radio is not bad at all, hence this separate chapter on shortwave broadcast mobile. But first some notes about interference.

CLEAN POWER

When it comes right down to it, there is no quick and easy method of installing sensitive electronic equipment—like radios—in automobiles. As sophisticated and specialized as much of today's dedicated automotive electronics might appear, a successful mobile installation remains a daunting task. Systematic and thoughtful procedures are essential if you want to minimize the noise and interference inherent in an automotive environment and get the most from your gear.

Believe it or not, one of the most important factors in optimizing your installation is the quality of electrical power you provide. There are three watchwords for providing mobile DC power: Clean, clean, and clean. The fact is raw automotive DC power can be an electronic jungle with high voltage spikes, AC ripple, noise, and power surges on the line to wreak havoc with the operation your sensitive mobile equipment. There are tried and proven methods of dealing with these problems, however, and we'll discuss them below. Another major entry on the list of noise sources is ignition interference and, as with providing clean DC power, careful and knowledgeable approaches to dealing with it yield the best results.

A GOOD MOBILE INSTALLATION BEGINS AT THE BATTERY

To minimize power-related problems, start your installation at the car's battery. A big lead-acid automobile battery is the best noise filter in a car and if the power supply to your equipment begins right here, you're in the best shape you can be. Your connection should use good quality loop-type lugs (not the open fork-type) attached to the bolts holding the terminal clamps. On batteries with top-mounted terminals this is easy: just install your lugs under the nuts holding the cable clamps. The situation is more difficult with side-terminal batteries. Here, you can get special side-terminal connectors with additional bolts for attaching accessories. Another alternative are replacement side-terminal connectors used when the originals have worn out; these have bolts on their clamps that hold the battery cables and your lugs can be anchored here.

Use heavy gauge insulated wire (#12 or heavier for loads of 10 amps or less) and solder all connections using a large soldering gun or other high-wattage iron; don't use small soldering pencils as they don't provide enough heat and poor solder joints can result. The idea is to ensure an excellent low-resistance path for both positive and ground leads to your equipment. At the battery, attach in-line fuse holders with fuses for the maximum power to be drawn by your gear to both positive and negative battery terminals; this will protect your equipment in case the battery's ground cable gets disconnected and your wiring suddenly becomes the ground line for the whole car.

Use string or a length of smaller gauge wire to carefully pre-measure the actual run from the battery to where your equipment will mount inside the car. For best results, try to route the wire as directly as possible to where it will penetrate the engine compartment firewall, noting places where you can tie down your final power cable.

The firewall feedthrough point can be an existing hole through which other wires are already installed. By taking some stiff wire and carefully probing around the edge of an existing bundle, you can determine the easiest feed-through point. Crawl under the dashboard and ensure your new wires will be clear of any mechanisms or other obstacles.

If you decide that drilling a new hole is the best method, make sure you obtain the right-sized rubber feed-through grommet to protect your wires from chafing on the edge of the newly drilled opening. Another alternative here is to use silicon-based caulking to coat the edge of the new hole and seal up the opening around your wire.

Cut sections of your heavy gauge wire to match the length of your pilot wire (plus about 20% more) and solder one end to the load side of each fuse holder. Leaving enough length for the fuse holders to reach their battery terminals, twist your heavy gauge wires together (one turn for about each 15 cm of length) so they form a cable, then wrap it tightly with good quality plastic electrical tape to just beyond the point where it will feed through the firewall. Truly hard core clean-power enthusiasts can use shielded pair cable or buy braided metal shield sleeving (available in various diameters) and slide it over their cable before applying the electrical tape; if you do this, connect the end of the braided metal shielding to the ground wire at the battery end and solder them both to the ground-side fuse holder.

Route your new cable to the firewall feedthrough point, using cable clamps or plastic ties to secure it along the way. Working the new cable through the hole in the firewall might require a little lubricant like a light oil or petroleum jelly. Remember to wipe it off later.

DC IN-LINE FILTERS

There are a variety of automotive in-line DC power filters on the market now and they are an excellent investment. Choose one with at least 20% greater current carrying capacity than your equipment will require (if you load these filters beyond their capacity, noise on the line can increase dramatically). Many commercial filters use both positive and ground leads and these are the ones to get (often the biggest interference problems in mobile applications are due to electronic noise on the vehicle's ground due to poor electrical connections between major body components). Find a good mounting location, cut your power cable wires to length and solder the filter into the line; the filter's output wires will now be your equipment connection.

At this point you face a decision. If you want your new power source to connect directly to your equipment, all you have to do is hook it up. If, however, you want to require that the ignition key be turned on (or in the "accessory" position), there is one more step. Go to an auto parts store and purchase a generic replacement headlight relay with contacts capable of handling enough current to power your equipment. To the relay coil, attach a wire that is only hot when the key is on (it usually powers the car's in-dash radio) so the relay closes its contacts when you turn on the key. Connect your new power lead to the relay's input terminal and the output to your equipment via a short length of leftover heavy gauge wire and you're done. Well, not exactly done...

INITIAL TEST

At this point you're ready to conduct your first test of how well your new power supply rejects the noise your car's electrical system produces. If the equipment you're installing is a radio, disconnect its antenna; if it's a tape deck or CD player, set it up with a blank tape or a CD at a quiet spot. Close the car's engine compartment, start the engine, and turn up the volume to see what comes over the speakers.

Among the most common problems is a high-pitched whine from the car's alternator that varies with engine RPM; second on the list is a staccato popping from the car's spark plugs, breaker points, or electronic ignition system. If you experience either of these, don't get discouraged; there are some other countermeasures we'll discuss below and this test isn't over yet.

It might sound like a cumbersome task but get hold of another car battery and put it on the floor next to your mobile equipment. Disconnect the equipment from your new power supply (both positive and ground wires), replace the leads with the external battery and repeat the test. If you still hear noise, it has to be radiated by the car's ignition/electrical system since it's certainly not coming from your external battery! Another reality is that you can never totally eliminate noise, regardless of what you do. Reconnect your equipment to the newly installed power wires and let's go on to some simple filtering.

FILTERING OTHER NOISE SOURCES

Examine your car's engine compartment carefully to locate the ignition coil, alternator, and voltage regulator. In newer vehicles the voltage regulator is often built into the alternator and the ignition coil is sometimes built into the distributor. No hard and fast guidelines here; changes in auto manufacturing are coming too quickly. You're looking for the output wires from the alternator and voltage regulator and the power input wire to the coil. These should be labelled but, if not, consult a reliable manual or service facility since you want to get this right the first time. You can purchase "filter" capacitors as a commercial product in many auto parts, electronics, and car stereo stores. These rugged devices are generally made with insulated wire, good quality installation lugs, built-in mounting clamps, and are excellent for this type of application.

An alternative is going to an electronic supply outlet and looking for some capacitors with a value of at least .2 microfarads (μF) and a voltage rating of at least 100 volts—200 is better. Make sure the capacitors are not the "electrolytic" type (these have a + terminal) since they don't work well in RF applications. Regardless of where you get them, these capacitors will serve to shunt high voltage spikes and transients on your DC line to ground.

On the engine's alternator (or generator), find the output terminal (usually labelled + or B+) and connect one of the capacitor's leads to it. Connect the capacitor's other lead (the metal casing, on commercial automotive filter capacitors) to a nearby ground point, minimizing the length of the wires. Do the same for the voltage regulator (if it's built into the alternator, your first capacitor will serve to filter both components). Find the ignition coil's power input and repeat the process with a another capacitor. In all cases, ensure you use good quality electrical lugs (soldered on in addition to being crimped) and insulate the leads in the event you use "electronics supply house" capacitors.

Other automotive noise sources include the windshield wiper motor, blower motor, electric fuel pump—in fact any electric motor that can operate when you're trying to listen to your equipment. On the market are devices known as "feed-through capacitors" which are designed to be installed in series with the equipment whose noise needs to be suppressed. Choose devices that meet or exceed the motor's current requirement and install them in the power supply line as close to the motor as possible.

AUTOMOTIVE RADIO FREQUENCY EMISSIONS

First, understand what's going on to produce the problem. When an electrical circuit is suddenly broken, energy tries to continue flowing across the gap in an attempt to maintain the current flow. The instantaneous energy build-up at the edges of the gap can produce high voltages and allow some of that energy to be emitted across a broad range of the electromagnetic spectrum. We can see this as a small spark but a large portion of this energy is in the radio frequencies and a radio receiver can detect it as wideband noise—or static.

Another noise source is the car's alternator and voltage regulator. Diodes—usually three of them in a "three-phase" system—rectify the alternator's AC voltage, leaving a small AC "ripple" riding on top of the resulting DC. If one or more of these diodes deteriorates or fails the ripple worsens, producing a ragged waveform rather than a smooth DC level and emitting small amounts of RF.

Automotive electrical systems are a nightmare when it comes to spurious RF emissions. Everything we've already discussed plus worn alternator slip rings, ignition breaker points opening, and spark plugs firing are all small, but intense radio transmitters. Metallic components in the car act as antennas, absorbing and re-radiating some of this RF energy. A portion of it, unfortunately, winds up being received or otherwise detected by your mobile electronics as interference. That's the bad news; the good news is these "car component" antennas tend to be fairly inefficient so shielding your electronics from their signals is not too difficult.

When it comes to protecting a mobile receiver from automotive radio frequency emissions, begin with one common-sense guideline: Get the radio's antenna as far away from the noise source(s) as possible. Many new car designers have already considered this and mounted the antenna on the rear of the vehicle rather than on the front fender next to the engine. If you are installing a new antenna keep this in mind and, if you have that option, exercise it. But if you're stuck with using the car's existing front-mounted antenna there are some straightforward and surprisingly effective countermeasures.

GROUNDING

Efficient grounding of major structural components should be your first priority. The car's hood or engine compartment cover is a good starting point since its electrical connection to the car's body is often less than perfect; if this is the case, the hood can be electrically "isolated" from the car's ground and act as an antenna to re-radiate—albeit inefficiently, as we've just discussed—the electrical system's radio emissions. But an inefficient antenna radiating noise can be a significant problem if your radio's antenna is mounted only a couple of centimeters away.

Make a short ground strap from heavy gauge wire or braided metal shielding and solder a lug to each end. Find a point where you can bolt one end to the car's body and the other to the engine compartment cover—preferably right next to where the antenna is mounted. Existing bolts are OK but make sure you strip the underlying paint down to bare, shiny metal before installing your ground strap. Afterward, you should be able to take an ohmmeter and measure less than 1 ohm of resistance between the car's body and any point on the hood.

Make sure the engine itself is ground-strapped to the body; this is usually done by the manufacturer but check for yourself and install your own if in doubt. Check other major structures: the trunk lid, the roof, the car's frame—all these components can serve as interference-transmitting "antennas." In a good mobile installation, there shouldn't be more than a few Ohms of resistance between any of the car's major metal components.

Another culprit can be the exhaust system. Often this whole assembly can be electrically isolated from the rest of the car since its mountings are usually rubber to damp vibrations and there is typically more oxidation on the metal parts since they get very hot. Oxidized metal connections typically have high resistance and can form crude semiconducting surfaces that re-radiate noise in the presence of RF energy. Here, you can go to a muffler shop and have them weld metal tabs drilled for your ground lugs to the muffler or other large metal exhaust system components. Later you can run short ground straps to a nearby spots on the frame.

SPARK PLUGS

Radio frequency emissions from spark plug wires can be a problem since they are difficult to eliminate. The fact is that most electronic ignition systems depend on spark plug wires being out in the open. When the engine is running the energy flowing from the distributor to the spark plugs produces a strong, rapidly changing electromagnetic field around the spark plug wires. Encasing them in material like braided metal shield sleeving can change their electrical characteristics, causing some of that energy to be dissipated before it gets to the spark plugs. This, in turn, can degrade the ignition system's efficiency since smooth electrical energy flow to the spark plugs has become increasingly critical in more modern automobiles.

What most auto manufacturers do to minimize spark plug RF emissions is to use "resistance" wiring or "resistor" spark plugs. This scheme inserts some electrical resistance—usually several thousand Ohms—into each spark plug circuit, rounding off the edges of the sharp electrical pulses the ignition system produces and reduces RF emissions. Commercially available after-market spark plug "noise suppressors" designed to be installed in each spark plug wire are usually resistors intended to convert normal wiring into "resistance" wiring.

From the aforementioned ignition system efficiency standpoint, it's generally not a good idea to combine resistor plugs and resistance wiring. If your car already uses either resistor plugs or resistance wiring, that's about all you can do. In stubborn cases, one alternative is to try and get special shielded spark plug wire. This material, though expensive and relatively hard to find, typically uses helically-wound conductor around a ferrite metal core to keep RF energy inside the wire.

THE END?

Don't feel frustrated if, after all your efforts, there is still some noise in your mobile system. The bottom line in any automotive installation is that you can never eliminate all the interference; you can only reduce it to acceptable levels. There are further steps you can take but the ones described here represent the most tried-and-true methods. Consult more technical literature for further guidance.

SW CAR RADIO MARKET

The myth that shortwave on car radios is widely available in Europe seems to persist. If you have about US $2000 you can indeed buy one top of the range model from each of Becker, Blaupunkt and even Sony that covers shortwave...of sorts. Coverage is limited mainly to the 49 meter band.

KENWOOD RZ-1

In 1989 the Japanese company of Kenwood launched their RZ-1 wide band receiver onto the European market. This could cover continuously from 500 kHz up to 905 Megahertz, i.e. medium wave, shortwave, VHF and way up beyond the UHF television bands in Europe. The price in Holland was 1499 Dutch Guilders at the time, which is US$832 at the current exchange rate.

Inside the foam packaging is a extremely light, compact, black colored receiver. It weighs 1500 grams and looks exactly like a car radio..in fact it will fit inside the standard hole drilled in the dashboard of most automobiles. If you do that though, you'll end up with a tinny muffled sound of less than one and a half watts. The small speaker is mounted on the TOP of the set, and sliding it into a car radio rack, of course, prevents the sound from coming out. Kenwood sell a small mobile speaker to mount on top of the dashboard, but if you are driving at anything above 20 kilometers an hour in a standard car, the noise of engine drowns out the 2 watts from the RZ-1.

MOBILE USE

If you are going to seriously use this receiver for mobile listening, you'll have to use a car hi-fi booster amplifier. That is no problem, because on the back of the RZ-1 are two phono connectors designed exactly for that purpose. Included with the instruction booklet are four plastic feet so if you want the receiver to sit on top of a table, either at home or maybe on board a boat, that can be done. You need to find a 12 volt DC power supply from somewhere. The unit consumes just under one ampere of current when it's on, so a mains power unit or a car battery are essential. You can forget about trying a bank of penlight cells - they would not last an hour.

The front of the receiver is neat and uncomplicated. There is a rotary tuning knob which has 24 steps per revolution, and which you can twist without banging your fingers on nearby control knobs. There is an on-off control, a squelch, and a row of buttons marked from 0 to 9. You use these keys to directly enter a chosen frequency. Each time you press a key the set beeps at you....the higher the number, the higher the beep. You can switch this beep off if you find it annoying.

PERFORMANCE BELOW 30 MHZ

Despite its triple-conversion circuitry (i.e. 45.75, 10.7 & 0.455 MHz), shortwave and medium wave performance on the RZ-1 is simply disastrous. Using a simple car telescopic antenna mounted on the rear of the vehicle we drove out into the countryside some 60 miles from any medium wave transmitters. Parked by the side of the road, we tapped in 9410 kHz onto a shortwave portable receiver costing about 200 US dollars....The BBC was coming in fairly weak, but readable. That was on a portable with the whip antenna poking out the window. We then listened to the same signal on the RZ-1. The easy to read white display showed 9410 kHz, but the signal was unusable, being masked by cross-modulation products.

The rotary tuning knob moves the received frequency in 5 kHz steps on medium and shortwave. There is no fine tuning control that you would expect on any communications receiver.

The dynamic selectivity of the filters is so mediocre that separation of stations 5 kHz apart is often impossible. We tuned the European medium wave band at dusk. Apart from two strong clear signals, the rest of the band was filled with "mush" and distortion. It is clear though, that in more remote areas of the globe where signal strengths are lower, the medium wave performance would be better. However, audio distortion (see later) remains a problem.

REAR VIEW RZ-1

CERTAIN CONTROLS NOT EASILY ACCESSIBLE

The RZ-1 may be called a wideband receiver, but you expect it to select stations one at a time, not all at once. There is an attenuator to reduce the signal level coming into the receiver. This helps slightly. But because they have put it at the back of the receiver, you cannot reach the attenuator switch if you put the RZ-1 in the dashboard of a car. The brochure indicates the receiver would be useful on a boat. As it has both AM, plus narrow and wideband FM modes you could certainly could use it to monitor local VHF traffic. But there is no single-sideband on the RZ-1, and undesirable second and third order harmonic products from the medium wave band obliterate most signals on the marine band above 1.8 MHz anyway.

For the shortwave broadcast and utility listener you can get significantly better performance than this on a communications receiver costing two thirds the price. However, the RZ-1 covers frequencies well beyond the shortwave band. If you connect a wideband discone antenna then you can start searching out the aircraft, utility and even FM broadcast stations. You can even listen to the FM in stereo via the external amplifier..but then any car radio will do that too.

100 memories are offered, together with four tuning rates of 5, 12.5, 20 and 25 kilohertz. These facilities are vital if you're scanning the huge area above 100 MHz for local signals. A unique feature is the ability to mark certain frequencies with up to 7 letters or numbers...so that when they come up you can actually see on the display your previously programmed identification.

Performance on VHF and UHF was considerably better than shortwave, though the sensitivity in the AM mode between 50 and 400 MHz is only fair. In the FM NARROW and WIDE modes sensitivity is quite adequate. The dynamic range is also better above 50 MHz. However, the RZ-1 did not match the specifications of either the FRG-9600 from Yaesu or the ICOM ICR-7000 which are the main competing wideband receivers on the market, albeit somewhat more expensive.

CONCLUSIONS

As far as the shortwave listener is concerned, the Kenwood RZ-1 is definitely NOT a good choice. The luxury of having a radio in the car that gives you the entire shortwave band with digital readout is not matched by the performance or the price. Any cheap car radio will produce less distortion on medium wave than the 14% total harmonic distortion that we measured on the RZ-1. You can also opt for a set with a cassette player too.

If you are looking for a VHF/UHF scanner, then check the competition. The RZ-1 clearly covers a wider part of the spectrum at the expense of performance. Kenwood have a justifiably good reputation in the field of communications receivers...their R-5000 model deserves to be popular. But we feel the RZ-1 is a disappointment. The price in the US is around US$500, whilst in Europe the receiver has gone from the market.

PHILIPS DC777

About six years ago, the Dutch Philips company launched their AC739 car radio stereo cassette player. This was much cheaper and offered two shortwave bands, in addition to medium wave and FM. One band covers 3.5 to 12 MHz, the other 12 - 22 MHz continuously. But the single-conversion set has only fair shortwave performance, and the backlash on the tuning knob makes tuning somewhat haphazard. Stations are crammed into a little dial space, and you could cause a nasty accident if you try to tune the receiver while driving at high speeds.....jamming and Morse interference leap out at you. That can startle if you are not prepared for it. The late 1987 catalogue incorrectly informed the customer that the unit is designed to pick up wavelengths of between 13 and 75 millimeters! The Spring 1988 Philips Nederland catalogue no longer listed the AC-739.

ENTER THE DC777

After many analogue attempts that were only mediocre, Philips has got it right this time with a no-nonsense car radio offering FM stereo, MW, LW, and shortwave. The coverage is between 3200 - 21850 kHz continuously. The radio also has a stereo auto reverse cassette mechanism (no Dolby noise reduction though), plus the ARI traffic info system used in German speaking parts of Europe. There is also a built-in clock timer, useful if you want to avoid missing a particular transmission.

There are two ways to avoid having the car radio stolen. For an extra US$40, Philips supply a removable slide. This allows you to remove the radio from the dashboard when you leave the car and take it with you (or hide it in the trunk). We found the quality of the connectors to be only fair.....constantly pushing and pulling the radio in and out of the housing caused some noticeable wear within 5 months.

The other solution is to mount the radio permanently in the dashboard and protect the radio with the security code system. This requires you to choose a security code. In ordinary use, you won't be asked for it again, but if the radio is disconnected from the car battery (as would be the case if it were removed from the car), it will ask for the code to be entered before the receiver or the cassette deck will function. The radio is software driven, so bypassing the code with a piece of wire and a soldering iron is futile. Its very important to make a note of the security code.....the radio will have to be returned to Philips if the radio asks for the code and you have forgotten it. This could be complicated and expensive if the radio is purchased outside Europe.

In some cases though, you may make a mistake when typing in the code. The radio gives you a second chance. If that is wrong then the radio blocks for 90 minutes. Don't turn the radio off! Leave the radio (and the ignition) on and wait. After the elapse time the radio will suddenly ask for the security code again. If you simply turn the radio off and try again in 90 minutes the radio will still be blocked. The instruction book is rather muddled in its English language explanation.

The DC777 is ideal for someone interested in hearing stronger international broadcasters while driving down the highway. The radio has a sharp 6 kHz filter, and an excellent AGC (important to combat the changes in signal level due to fading) What appears to be a large tuning knob at first (in the bottom right hand corner) is a volume control, which also doubles as a fader, tone, and balance control. Tuning is achieved through a lighted direct entry keypad which folds out at 45 degrees when you press a button marked "OPEN". The two rows of six keys have a soft spongy feel which some might find rather clumsy to use. In any event, it would be unwise to try and tap in frequencies and try to concentrate on the road.

All is not lost however. Two buttons, UP and DOWN set the receiver scanning the dial for strong stations. You can also elect to manually tune the receiver by pressing both buttons at once. After that, pressing UP or DOWN shifts the frequency up or down by 1 kHz...perfectly adequate for broadcast listening.

The radio also has a row of 5 memory buttons. In fact on shortwave, by introducing a "banking" system, the radio can store up to 20 favorite shortwave channels. On FM you also have the option of activating an automatic storage system, which scans the FM and picks the 5 strongest signals and stores them under the memory buttons..handy on long distance drives.

Shortwave in the car is possible by connecting a standard portable (such as a Sony 2001D) to the car antenna. The problem is that the volume produced is often drowned out by the engine noise. The DC777 will drive up to 4 speakers with 7 watts RMS per speaker....indeed the audio fidelity of the shortwave reception is quite astounding.

The radio comes complete with a handbook about shortwave, but the station information, as is so often the case, is seriously out of date. The radio, manufactured in Singapore, fits the standard DIN slot in European cars. If the radio doesn't fit your car, then a mounting bracket of some kind will be needed.

CONCLUSION

Philips have come up with a down-to-earth car radio with good shortwave capability. Whilst the tuning may be a bit fiddly for some, the quality of reception is good. It is important to mount the antenna well away from cable associated with the car's ignition system, or the on-board computer now used in many vehicles. A bit of experimentation before final installation is recommended.

The price of the set varies enormously. In The Netherlands it first appeared at 1295 Dutch Guilders. This puts the radio into the mid-price range for car radio/cassette players. But Universal Radio in Ohio USA has the radio listed for just under US$400, considerably cheaper.

OTHER SW CAR RADIO OPTIONS

XRU 882 RDS

At the Friedrichshaven ham radio fair in 1992, Sony showed a new car radio with the type number XRU 882 RDS. It costs in the region of US$1000, and boasts medium wave and FM reception plus shortwave. There's also a high quality cassette player built-in, plus an option to connect a compact disk player. Just what is meant by shortwave though wasn't clear at Friedrichshaven. The technical specifications were vague when it came to shortwave coverage. The radio also offers a two antenna FM diversity reception system, which reduces the effects of fading. Further enquiries show that the radio is in fact only equipped with the 49 meter band.

GRUNDIG CAR RADIOS AND SHORTWAVE

In the last couple of years Grundig Germany has added one budget and two top of the line car radios with shortwave capability. The WKC4805 is the cheapest of the range offering an excellent cassette deck and a MW/LW/FM/ and SW radio. The problem is that the radio does not receive the full shortwave range...only the 19, 25, 31, 41 and 49 meter bands. That is strange bearing in mind that Radio Austria International, one of the most pleasant German speaking international broadcasters, booms in across Europe on 13730 kHz. The lack of coverage in the 22 meter band, plus 16 & 13 meters is a shame. The power of the amplifier is just 5 watts per channel which is on the low side for a car, but the price of DM500 is not bad.

WKC 4805 RS

The two top-of-the-line Grundig car radios offer the possibility to control a CD player as well as the built-in cassette deck. The power amplifier delivers a healthy 4 x 20 watts RMS in each case. But curiously the shortwave coverage of both these sets is more restricted....49, 41, 31 and 25 meter bands only. A clear case again of "almost but not quite". We hope that Grundig will follow the Philips example, even if it is for a specialized market.

WKC 4870 RDSC

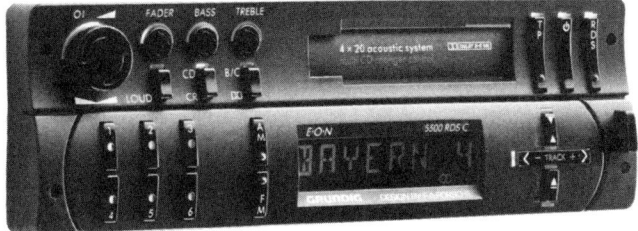

WKC 5500 RDS-C

TOPY-200 SHORTWAVE CONVERTER

Just when we thought the shortwave car converter market was finished, we managed to purchase an interesting converter through the technical services department of the German shortwave club ADDX.

The converter was originally designed by a Korean company for a Turkish businessman living in Central Germany. He thought there would be a big market for such converters so that Turkish citizens living in Germany could receive programs from the TRT in Ankara. 1000 examples were made but the company went bankrupt as the converters arrived in Germany. The ADDX has been able to buy some of the bankrupt stock and sell

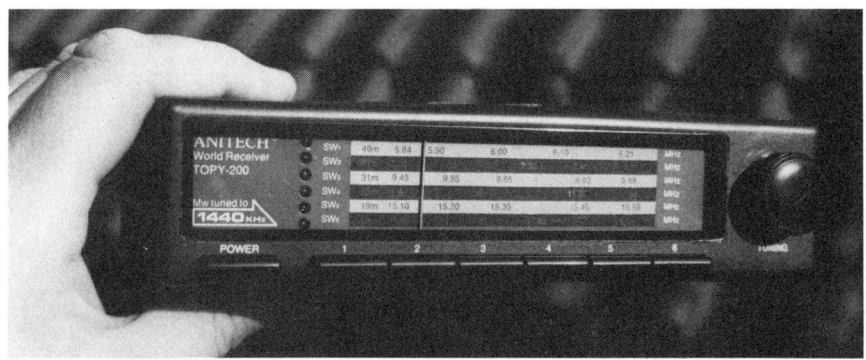

it to members for 100 German Marks. The converter has some limitations, but for that price you can't really grumble.

You simply mount the converter between the car antenna and the input of the car radio. The old car radio converters from Grundig and Becker used the main tuning knob on your car radio to tune in shortwave. With the move to digitally tuned sets that became impossible. Fortunately the TOPY converter works the other way round. You set the ordinary car radio to 1440 kHz AM, and tune the converter instead. We found the TOPY to be incredibly sensitive. It suffers more from splatter than the DC777, but for the price paid gave remarkably good results. Power consumption is quite low - around 200 mA. More information from Charly Hardt, Edelhoffstrasse 70, D-5630 Remscheid, Germany. Tel:+49 2191 80598.

Finally we think that Lowe Electronics would be wise to release a mounting bracket so that the new HF-150 can be mounted in the car, although some form of booster amplifier would also be a good idea.

CHAPTER 10
RADIOS AND COMPUTERS

The use of personal computers (PC's) in connection with international broadcasting is rapidly increasing. A few years ago the International Telecommunication Union in Geneva announced the availability of the Tentative High Frequency Broadcasting Schedule in a computer-readable format. Hitherto, that publication, which contains the registered shortwave frequencies of ITU member countries, has only been available as a thick book published four times a year.

A few international broadcasters already make their schedules available to computer owners through bulletin boards, and individual listeners often use this method to pass on schedules and interesting loggings for others to share.

To use a computer bulletin board, you need a device called a modem to connect your computer with the telephone system. You will also need a suitable communications software program to control what happens while you are actually connected to the bulletin board.

If this sounds too complicated, there are other ways you can get computer-readable information about international broadcasting. Two easy-to-use programs for IBM and compatible computers enable you to maintain an active database of interesting broadcasts. One is aimed at the North American market, the other at Europe.

SHORTWAVE BROADCAST SCHEDULES

Four years ago, the Shortwave Broadcast Schedules database program from TRS Consultants was one of the first WRTH Industry Award winners. Since then, the program has been further improved, and in our opinion still sets the standard in its category.

The menu-driven program is very easy to use, and is supplied with the latest updated data files containing the transmission times and frequencies of all English language international broadcasts audible in North America, and a separate database of DX and media programs.

A problem for many new listeners to international broadcasters is working out the difference between local time and UTC. A major enhancement to the original program now enables the user to set the time difference when installing the program (it can be altered to take account of Daylight Saving Time). Using the new option 'C' for 'Current' will make the program display only those broadcasts on the air at that moment. It does this by reading the time from the computer's clock and then making the necessary adjustment for UTC, as set by the user.

Of course, a program such as this is only as good as the data it contains. TRS Consultants go to great lengths to ensure that users have the opportunity to keep the data current. Those with a modem can call the Pinelands Bulletin Board and download the latest data files. It is also possible to subscribe to updates on floppy disk. Members of the North American Shortwave Association receive printed updates free at regular intervals in their club journal.

It is also simple to update the information yourself using option 'U' on the menu. But, of course, you are then responsible for the accuracy of the information ! Unless you are a prolific listener and have access to a lot of current information, we would suggest that you would do better to rely on the official updates in whichever form is most convenient.

There is one aspect of the program that we find a bit clumsy, but that is not the fault of the author. When you install it for the first time, the install procedure automatically indexes the data files. However, because the program uses the industry standard dBase file formats and procedures, it is also limited by the shortcomings of that program. When you load updated data files, you must manually select the option to re-index the databases. If you don't do this, the program isn't clever enough to know that something has changed, and will probably display information about the wrong station or transmission ! An inexperienced user would probably not realize the cause of the problem.

The price for the program is US$30, and then US$35 for an annual subscription to the schedules. An up-to-date listing is kept on the Pinelands bulletin board which can be accessed at speeds up to 14,400 Baud. The phone number for the bulletin board is +1 609 859 1910. Further information available from: TRS Consultants, P.O. Box 2275, Vincentown, NJ 08088-2275, USA. Tel: +1 609-859-2447. Fax: +1 609-859-3226. MCI Mail: 244-6376. Telex: 6502446376MCI UW.

THE RADIO LISTENER'S INTERNATIONAL PC DATABASE

While the program from TRS Consultants concentrates on times and frequencies, and restricts itself to broadcasts in English, a rather different approach has been taken by a Norwegian company called Elektrokonsult. Their product, with one of the longest names we have seen for a software program, sets itself the rather ambitious task of providing data on different categories of programs in more than a dozen European languages.

Using a menu system, you choose a language and then a program category (such as sport or music), plus options such as target area and days of the week, and the program will search for and display the names of all programs that match the specified criteria, giving their transmission times and frequencies.

The program certainly works in the manner described in the 60 page User's Manual, but the problem with the version under review is that the amount of supplied data is extremely limited, and randomly choosing various combinations of language and program type revealed no matching data at all!

To be fair, Elektrokonsult do admit this, and they say they are expanding the database all the time, but because of this limitation it is difficult to judge how accurate or current the information supplied in later releases will be. The latest updates are supplied when you order the program, but after that you can purchase single updates or take an annual subscription (four updates).

We can, however, comment on the user interface. We are pleased to see that someone has written a database program that allows the user to select a shortwave broadcast according to its subject matter. This should broaden its appeal, so it is all the more disappointing that accessibility to the information is limited by an interface that has more in common with DXing than program listening.

Because of the large number of possible permutations of language, program category and other criteria, the author has chosen to use a system of abbreviations, not all of which were immediately obvious to us. There seems to be plenty of unused space on the screen, and we think that the program would be greatly improved for the novice by expanding many of these abbreviations into something more easily understood.

We also wonder whether the program is trying to do too much. Assuming that sufficient data will eventually be included with the program, does anyone other than a professional linguist really need details of programs in a dozen languages? It might be better to make the language modules optional, so that the customer just pays for the data he really wants.

As with the TRS Consultants program, it is possible to update and expand the data files yourself, so a keen listener could build up a personalized database matching his own individual needs. Here, too, because the program uses the dBase file format, manual re-indexing is required whenever you change data to ensure that the program knows where to find everything.

In conclusion, then, we feel that the Radio Listener's International PC Database is on the right lines, but needs further development. We hope to have the opportunity to review a later release of the program in a future edition of the WRTH.

Available from "Elektrokonsult AS, P.O.Box 846, N-3002 Drammen, Norway. Basic Program US$49 + US$5 postage. Single update to data files: US$25 + US$5 postage. Annual subscription for four updates: US$49 + US$5 for postage. Specify whether a 5.25 inch (360k) or 3.5 inch (720k) disk is required.

PUBLIC DOMAIN SOFTWARE FOR THE MACINTOSH

DX Window: Creates on screen an azimuthal equidistant projection (great circle) world map centered on your location, with day/night terminator. Engineering Systems, Inc., PO Box 939, Vienna, VA 22183 USA. $39.95

Skycom 1.1: Enter solar flux and get propagation predictions to desired areas of the world.Engineering Systems, Inc., PO Box 939, Vienna, VA 22183 USA. $39.95 .($59.95 with Skycom 1.5)

Skycom 1.5: Provides sunlight status at both ends of the path; MUF, F0F2, and FOT frequencies; S/N ratio of the link, and other information. Engineering Systems, Inc., PO Box 939, Vienna, VA 22183 USA. $39.95 ($59.95 with Skycom 1.1)

Sun Clock: A useful color Desk accessory. Displays a map of the world with day and night areas. MLT Software, PO Box 98041, 6325 SW Capitol Highway, Portland, OR 97201, USA. Cost is $17.00

DX HELPER / SATELLITE PRO

MacTrak Software (Box 1590, Port Orchard WA 98366 USA) has come up with two excellent programs for the communications enthusiast. DX Helper (US$39.95) works out beam headings and distance, the current sun position, gray-line, and draws a great circle map based on any listening location. Some of the features (e.g. DXCC Country List) are of more interest to amateur radio operators. Another program entitled "Satellite Pro" (US$99.95) permits automatic tracking of amateur and meteorological satellites, even to the point of driving the elevation-azimuth controller. The programs are distributed by Antennas West, 1500 North 150 West, P.O. Box 50062, Provo UT 84605 USA. Tel: +1 805 373 8425.

SW NAVIGATOR

This Mac program won the WRTH 1989 Industry award. Author Jim Frimell has now released version 2.0 with a lot of major improvements. These include better quality ID signals, and easier sorting of files. This is still the best "program" guide program on the market, and now drives the JRC NRD-535, the Kenwood R-5000 and the Drake R8. The program costs US$99.00 in the US. Further information from SW Navigator, 232 Squaw Creek Road, Willowpark, TX 76087 USA.

EXAMPLE OF SW NAVIGATOR FILE

BANDVIEW 1.50I FOR THE PC

This professional controller and station logging system has been developed for three receivers, the JRC NRD-535, the NRD-525, and the Kenwood R-5000. The software package is specific to one of the receivers, in our test case the NRD-525. Apart from the receiver (fitted with the appropriate computer interface if necessary), an IBM compatible computer with an RS232 serial communications adaptor. Apart from that, the software does not require third party products such as MS Windows.

Once up and running, you can control most of the receiver functions such as frequency, mode, bandwidth selection, attenuation, AGC constant, and the memories from the computer keyboard. Other programs do this too, but the key to Bandview is the link with a loggings database. You can build this database up yourself, or import it from bulletin boards (e.g. Tom Sundstrom's excellent service of schedules in English.)

The screen gives you an instant overview of the receiver's current status. It also displays the relevant portion of the electronic logbook, highlighting when the schedule of the logged station matches what you're listening to at that moment. For instance, if you tune to 15410 kHz around 17 hours UTC, one of the possible signals might be VOA's African Service in English. Bandview is ideal for bandscanners, because it is possible to link various broadcast bands together. So having reached the top end of the 16 meter band, the receiver can be set to spring to the lower end of the 13 meter band. This program is designed for shortwave broadcast and utility listening. It is not a ham radio program that has been converted to accommodate the listening side. The program requires the minimum of hardware to get it running. It's possible to set the computer for unattended recording (i.e. 11835 kHz at 0030, followed by 6195 at 0130 UTC).You simply set an autoflag using a single stroke command. Bandview cannot directly control a tape-recorder for unattended recording in the versions for the R-5000 and NRD-525. In the NRD-535, the accessory socket is software controlled. The 535 version also incorporates a spectrum scan 15 kHz either side of the selected frequency.

It is easy to add a new log as you find stations of interest. You can also export the logs in a variety of formats. If the station is new, there's an option to automatically generate a QSL letter, using the logging details. The program allows the use of the SINPFEMO code (an expanded version of the SIO code), although whether all radio stations will understand this code is questionable. Codes are subjective, so apart from signal, interference, and overall merit ratings, a few lines describing the problem says more than a subjective code. The screen display is best in color, although it will run in monochrome. The clock has data about world times built-in.

In short this is an extremely versatile program, its main advantage being the link between current listening and the logbook (up to 1000 entries open at any one time). It costs US$120 including shipping in North America. That price makes the program excellent value. A passive demo of the program is available for US$5.00. More details from Tom Kashuba, 2000 Commonwealth Avenue, Suite 1407, Boston, MA 02135 USA. Tel: +1 617 782 6660 (14-20 UTC).

MACRATT FOR THE PAKRATT 232.

Advanced Electronic Applications has been producing the PK-232MBX data controller for some years. As software revisions have been made, so existing owners have been able to upgrade. You simply connect the PK-232 between a shortwave communications receiver and a computer (IBM or Macintosh) to decode radio-teletype, morse, AMTOR, FAX and packet radio.

The hardware side of the unit is well-made, and a row of LEDs shows the current status and mode of the unit. Operation for more types of utility signal is straightforward, though you would not gain that impression from the manual. AEA primarily supply the amateur radio market, and thus assume that users understand the jargon. As a result, a growing market of listener only users is being confused.

At a recommended price of US$59.95 the MacRatt with FAX Terminal program (plus the necessary computer cable) is good value, but it needs some upgrading bearing in mind the capabilities of the Mac computers running under System 7. The menus could be a lot easier to use, with, for example, RTTY shifts instantly shown on the screen. The PK-232 has been an excellent piece of hardware for several years. The software now needs to enter the 90's. More information from Advanced Electronic Applications (AEA), PO Box C-2160, Lynnwood, WA 98036 USA.

SEEKER-PC FOR THE KENWOOD R-5000

This program is an extension of an control program originally designed for the Commodore 64. In developing a PC version the software designer has gone back to the drawing board, incorporating the best features of the old program into something quite new. An IBM-PC with 512k RAM and DOS 3.1 or higher, one available serial port, and a game port are needed. The R-5000 needs to be fitted with the IC-10 kit, and a jumper wire must be installed inside the set. The instruction manual gives very precise instructions as to how this should be done, but also warns that it may invalidate the warranty. In practice it took us about 15 minutes to make the modification.

Seeker is ideal for listeners following major news events. As you find stations you can quickly log all the details into a database. Then, the radio can be set up to record important newscasts, starting and stopping a tape recorder at the right moment. The ability to analyze the received signal strength allows the program to judge whether alternative frequencies would be best, and switch to them. This is ideal for unattended recording. Sequential scanning of chosen frequencies or an entire band is possible, again unattended if need be. You can set the minimum required signal strength. Finally, the program can be used to dump useful frequencies into the R-5000's memory when the radio is to be used of DXpeditions. The use of pull-down menus and hot keys greatly enhances the user-friendliness of the program.

There are clearly users who want the sophisticated scanning and tape-recorder control functions provided by Seeker-PC. The program only works with the R-5000, and currently no development is taking place on versions for other receivers. The entire Seeker PC system, including documentation, software, interface, and power supply costs US$392.50. This puts the program into the semi-professional sphere, since it is half the price of the receiver it's designed to operate. Further information from AF Systems, P.O. Box 9145, Waukegan, IL 60079-9145, USA. Tel: +1 708 623 4744.

INTERVAL SIGNALS VERSION 1.21

One of the easiest ways to identify an international broadcast station is by its interval signal. Most stations play a short tune or a few bars from a longer piece of music, before the start of the transmission and between programs in different languages. In the past, cassette tapes have been available containing recordings of the various interval signals, but none have been issued recently, and there have been a lot of changes.

All personal computers have some kind of loudspeaker. In the case of the IBM PC and compatibles, the speaker is not exactly hi-fi, but is capable of producing a recognizable tune using quite simple software. We used a program called Pianoman to produce a reasonable rendition of the Radio Netherlands interval signal in only a few minutes.

But the average shortwave listener is more interested in hearing the interval signals which he or she doesn't already know. For that purpose,

Mark Fine has produced a program called Interval Signals, which will run on any IBM compatible PC and does not demand much in the way of disk space or memory.

Using the program couldn't be easier. Typing "IS" on the command line loads the program, which briefly displays a simulation of a digital frequency dial and keypad while it reads in the data file that contains the music. It then proceeds to display a simple text based screen giving a menu of stations in alphabetical order. The chosen station is selected by using the cursor keys on the keyboard, and pressing the return key plays the appropriate melody.

The program was apparently written to be compatible with the old CGA (Color Graphics Adaptor) display standard, and using a Hercules monochrome display the selection "bar" is in fact just a broken line underlining the selected station - however, we had no problem with this, and it did not hinder our use of the program.

As supplied, the program defaults to a playback speed which was optimized for the slowest computers, having a clock speed of 4.77 MHz. This can be re-set to any value up to 35 MHz, and the program automatically stores the selected value as the new default. The author recommends choosing a value equivalent to the clock speed of the computer in use. However, we found that using a 12.5 MHz machine, the speed had to be set to 25 to make the tunes play at a sensible tempo.

As for the actual tunes, the author has chosen to leave in some old interval signals - for example Radio Berlin International, which already closed before the end of 1990, and Radio Tirana's "With Pickaxe and Rifle" which was dropped during 1991. Radio Bucharest's new interval signal is designated "Radio Bucharest 1" while the old one is still there as "Radio Bucharest 2". On the other hand, Radio Sweden's old interval signal is "Radio Sweden 1" while the new one is "Radio Sweden 2". We think that the obsolete data should have been omitted in order not to confuse newcomers to international listening.

We noted some musical inaccuracies in some of the renditions - not serious enough to make them useless, but rather irritating to the musical ears of someone who knows the tunes already. Inexplicably, the third note of the Blue Danube waltz of Radio Austria International is quite different from the one Strauss wrote ! This may be due to the author misreading the music from a printed source rather than listening directly to the broadcasts.

Despite these criticisms, the program is good value and worth having, if only because it is almost free of charge - the author requests a nominal donation of US$5 when you order a copy from anywhere in the world, which goes to cover his costs plus postage.

The program comes complete with on disk documentation, and details of how to keep in touch with the author. It can be ordered from : Mark J. Fine, c/o FineWare, Inc, 11252 Cardinal Drive, Remington, VA 22734-9684, USA.

INTERVAL SIGNALS ON LINE

Another source of interval signals for IBM compatible computers is an ongoing project by American-based DXer Richard Ure§a. At the time of writing (August 1992) a total of 65 interval signals is available, and the list is growing. The programs can either be run using BASIC (a version of which is supplied with every copy of Microsoft's DOS operating system), or there is a compiled version which runs directly from the DOS command line. Access to these files is possible through computer bulletin boards which belong to the FidoNet system (see details at the end of this chapter).

SHORT WAVE LOG 1.10

For some years, a number of DX clubs have used computers to assist in the preparation of their monthly bulletin. Recently, several programs have been produced which allow the individual short wave listener to use a computerized logging system instead of a paper log book. The one we review here was written for IBM and compatible personal computers.

Unlike the Interval Signals program, this one will certainly not run on a minimal system. In fact, page three of the User's Guide states that the program "places rather heavy demands on the system". In practice, this translates to a minimum requirement of at least 450k of free memory and 2.5Mb of free Hard Disk Space. This may rule out its use by some hobbyists - machines with limited memory and no Hard Disk may not be manufactured any longer, but there are a still a lot of them in use !

For those with a more advanced system, however, the program has been written to take advantage of expanded and extended memory, and also of the international language support provided with recent versions of the computer operating system (DOS). It also supports multitasking under DESQview. In short, this is a program written to professional standards.

Before using the program, it is necessary to tell it which time zone you are in, enabling the program to display the correct UTC time. Details of how to do this are in the User's Guide under the rather intimidating heading of "Environment Variables". The user is then told to enter some information on the command line before the program will even run, but the manual only gives a single example and assumes the user will know what to enter. We would like to see this aspect of the program improved in future releases.

The program is supplied with a limited database of international broadcast schedules. The version we tested had 400 records with data for the summer period. This is accessed from the Freqlist menu. Presumably the data will be updated twice a year so that new users receive accurate information when they buy the program. However, it is very easy for the user to update and add to the supplied data. This implies that experienced shortwave listeners will be able to do more with the program than novices, simply because they are likely to have more information to add !

A separate database is used for the user's own logs, and is accessed from the Logfile menu. Because the program uses industry standard relational database software, it is possible for the user to check the details in the log file against the freqlist database. In other words the program can help identify a station by indicating a likely match. For the professional user, this facility is potentially very useful - clearly a much larger database would be needed to use it effectively. For the less experienced hobby user, however, there is a danger of jumping to the wrong conclusion based on what the program suggests, especially when the data is incomplete.

A different file contains the addresses of the stations included in the schedule database. This proves very useful when used with another major feature of the program - the QSL writer, which makes writing reception reports easy. Selecting this option produces a well designed data entry screen. There is a choice of four languages for the output - English, German, Spanish, and Swedish. We are puzzled at the inclusion of Swedish, as most Swedish DXers write their reports in English. French would have been much more useful for tropical bands DXers ! However, further language modules are promised for the future.

If the station to which you are writing is in the database already, you only have to select its name from the menu, and the program fills in the rest of the address automatically - a very useful feature ! When you reach the point of entering program details, a menu pops up on the screen automatically with all the major items you might want to mention - everything from interval signal, sign on, news, features, music right to the sign off. Selecting one of these items immediately places it on the report form, and you then add your own details. We liked this feature - it encourages good reporting discipline, because while it reminds you what you should include, it still requires you to enter details of your own to make the report authentic.

We have two criticisms of the report feature in its present form. The supplied text editor does not have automatic word-wrap, although the manual explains how to use a different text editor instead of the one provided. Also, like so many programs of American origin it requires the date to be entered numerically in U.S.format (MM/DD/YY) and will give an error message if the user tries to enter the information any other way. Thus December 31st 1991 has to be entered as 12/31/91 - entering 31/12/91 doesn't work. We would like to see an option to customize the format - a European user entering 04/01/92 on the first of April will find that the program accepts the data, but prints "January 4th" on the report. This can be very irritating since the manual does not explain the problem.

When the report has been completed, pressing ctrl+enter on the keyboard prompts you for a file name, and the completed report is saved as a neatly formatted text file, which you can then edit if you want to before printing it out and sending it. Compiling a report this way is certainly quicker and more efficient than writing one by hand - and the station will be able to read it clearly !

In summary, this is a very professional product, written using the most up to date programming techniques. For an experienced user of an IBM compatible PC, we think it is an excellent productivity tool, but we don't recommend it for novices. It has clearly been designed for the American market, and needs some further development to optimize its usefulness outside North America. Even so, at a price of only $20, its cost/performance ratio is outstanding. The program is shareware, and is available for download on some computer bulletin boards that specialize in radio. Otherwise, you can contact the company direct at the following address: Lee Consulting, P.O.Box 71303, Pittsburgh, PA 15213, USA.

OTHER RADIO-RELATED SOFTWARE

The UK-based Public Domain and Shareware Library (PDSL) has an impressive range of IBM PC software for radio hobbyists. This includes such categories as logging and QSL programs, propagation and receiver control software. Programs are available on disk or on line from PDSL's multi-line computer bulletin board. Non-members can purchase disks and access a limited number of files on the bulletin board, while members can get cheaper disks and full access to the BBS. A copy of the current PDSL catalogue which includes membership details is available for UK£2.00 from the PDSL, Winscombe House, Beacon Road, Crowborough, Sussex TN6 1UL, England. Tel : +44 (892) 663298. Fax : +44 (892) 667473. The Bulletin Board number, from which file lists and membership details can also be obtained, is +44 (892) 661149. Modems should be set for 8-N-1.

COMPUTER BULLETIN BOARDS

Apart from radio programs such as Media Network on Radio Netherlands, one of the fastest ways to get the latest information on shortwave broadcasting and equipment is via a computer bulletin board. In Europe and North America, a good personal computer and modem can now be obtained for less than the price of many of the tabletop receivers mentioned in this publication.

There are two main types of Bulletin Board System (BBS). The first consists of the subscription-based professional services such as Compuserve (TM) and GEnie (TM). The other type consists of boards run by enthusiasts for either a modest subscription or completely free of charge. Many of these boards link up with each other in a de facto network which permits them to exchange messages with each other, and thus enables a user of one board to contact a user of another board on the opposite side of the world.

The oldest of the big international networks is known as FidoNet, and there are an estimated 10,000 computers worldwide connected to it. The messages are organized by subjects into Conferences, or "Echos" as they are known to FidoNet users. The Shortwave Echo has grown rapidly to become one of the largest on the network, generating approximately 100 messages a day. The way these messages are sent around the world is quite remarkable - computers link up in a giant hub and spoke system rather like the route system of an international airline.

Example of some of the files to be downloaded on a typical SW bulletin board

File	Size/Date	Description
GEO_BULL ZIP	18540 03-28-92	Geoclock RBBS' Bulletins, Here's all the information off Ahlgren's BBS on the BBS, Geoclock and maps.
ARABIA ZIP	1280 03-29-92	Add Arabian "country" borders. For Geoclock 4.3 or higher.
BRITAIN2 ZIP	1821 03-29-92	Add borders between England/Wales/Scotland. For Geoclock 4.3 or higher.
GDATA1-5 ZIP	12795 03-29-92	Geo-data for maps 1-5.
GE-ICONS ZIP	18102 03-29-92	MSWin icons for Geoclock.
MAP9000 ZIP	10315 03-29-92	US Equi-azimuth map.
SVGA1-2 ZIP	69422 03-29-92	World & 48 states SVGA maps.
Z92RFE ZIP	3285 03-27-92	Radio Free Europe Z92 schedule.
NET0392 ZIP	9216 03-31-92	ANARC SWL Net Logs March '92
NET1Q2 ZIP	30720 03-31-92	ANARC Net logs RECAP Jan-Mar '92
NRD0392 ZIP	5120 03-31-92	ANARC Net Logs for Whiteside Mar '92
NETUTE12 ZIP	8192 03-31-92	ANARC Net Logs UTE EXTRACT Jan-Mar '92
INTRO_RT TXT	23373 04-02-92	Intro to GEnie's Radio Roundtable. Come join the SWL gang on GEnie. The telecommunications software is in the TELE download directory.
SCDX2153 TXT	14199 04-07-92	Sweden Calling DX'ers #2153 04/07/92
DAIL0392 ZIP	48014 04-11-92	Daily Solar Geophysical Data: Mar 92.
SW_MSW3 ZIP	2598 04-11-92	Eng Lang SWBC Schedules: MSWin3 PIF & Icons. Here's a way to run the program under Microsoft Windows 3.00a or 3.1.
CIS9204A TXT	11910 04-14-92	Commonweath of Indep States' schedules.
BBCZ92 ZIP	13814 04-17-92	BBC the spring-summer schedule
BBCPRO 492	15366 04-17-92	BBC World Service revised program schedule
GEOCK444 ZIP	181214 04-18-92	Ahlgren's GEOCLK v 4.44 rel 4/11/92. World Time/Maps. Requires EGA/VGA/SVGA. Minor upgrade to v 4.40.
MAPS3-5 ZIP	48501 04-18-92	GEOCLK maps 3-5 (map 5 has been updated.)
ANARC MAR	11524 04-19-42	ANARC Newsletter: March 1992.
535_AGC TXT	4894 04-19-92	NRD-535 AGC modification to decay rate.
PHON_ANT TXT	5939 04-19-92	Turn your telephone line into an antenna.
M6000 TXT	11937 04-19-92	Control M6000 & M7000 with PC.
SAC TXT	5290 04-19-92	Strategic Air Command HF frequencies.
CUSTOMS TXT	3675 04-19-92	US Customs HF frequencies.
RCI_Z92 TXT	12160 04-19-92	Radio Canada Int'l schedule eff Mar 29.
RN_Z92 TXT	8705 04-19-92	Radio Netherlands schedule eff Mar 29.
WJCR TXT	2582 04-20-92	Technical info on WJCR/KY (7490 kHz).

The end product of all this frantic on line activity is rather like an on-line DX club, where people exchange DX tips, experiences of using various bits of equipment, anecdotes and a considerable amount of light-hearted banter. Regular callers use a program called an offline reader which checks the BBS for new messages, sorts them and compresses them into a file which is then downloaded by the caller. The user reads and optionally replies to the messages while off line, and the program prepares a reply file which is uploaded at the time of the next call.

Participants in the Shortwave Echo cover all categories, from the young hobby listener to professionals such as the Voice of America's frequency manager. Andy Sennitt, editor of the World Radio TV Handbook, is an enthusiastic supporter of the Shortwave Echo. Andy says "it's an ideal way to keep in touch with the views of WRTH readers, learn new information and generally find out what's going on beyond the confines of the editorial office".

Not all the bulletin boards connected to FidoNet necessarily carry the Shortwave Echo. Each sysop selects a range of Echo subjects which best reflect the interests of the board's regular callers. But most sysops are only to happy to try and get a new Echo on request, which sometimes takes a few weeks to organize due to the voluntary nature of the FidoNet operation.

Below are the numbers of some bulletin boards in North America, Europe and the Pacific which are known to carry the FidoNet Shortwave Echo. If you use a BBS which is not currently carrying it, ask your sysop to contact his national or regional FidoNet coordinator.

This is a list of 338 bulletin boards known to carry the Shortwave Echo service as of September 1992. The list was compiled by Richard Ureña. NOTE: Some nodes operate part-time only. ALWAYS try a voice call first, before attempting to communicate via modem.

Fido Node	Board Name	Location	Phone Number	Maximum Baud Rate
1:1/110	Macintosh Help	Somerville MA	1-617-625-0381	9600
1:102/128	Ursa Major BBS	Manhattan Beach CA	1-310-545-7216	9600
1:102/138	Long Island RB	Los Angeles CA	1-310-370-4113	9600
1:102/420	Target Range BBS	Paramount CA	1-310-634-8993	9600
1:103/100	Orange Co West	Garden Grove CA	1-714-638-2298	9600
1:103/111	Amateur Radio CBCS	Garden Grove CA	1-714-633-2963	2400
1:103/148	Attitude of Gratitude	Buena Park CA	1-714-527-0811	2400
1:125/28	Coconino County	San Francisco CA	1-415-861-0311	9600
1:161/210	Politics & Religion	Concord CA	1-510-682-5179	2400
1:202/114	The Chief's Mess	San Diego CA	1-619-469-1354	9600
1:202/302	Gandalf's	San Diego CA	1-619-466-9505	9600
1:202/702	The SANTEE Experiment	Santee CA	1-619-562-8758	9600
1:202/711	PRI Wildcat! BBS	San Diego CA	1-619-278-7361	9600
1:202/719	Interface	San Diego CA	1-619-297-7733	2400
1:202/731	Hertzian Intercept	Mira Mesa CA	1-619-578-9247	2400
1:202/1201	OS/2 Desktop	Escondido CA	1-619-743-2511	9600
1:203/333	The Rancho Connection	Citrus Heights CA	1-916-722-5615	9600
1:205/40	Fresno Area Amiga eXchange	Fresno CA	1-209-226-7162	9600
1:205/35	West-Net 1	Fresno CA	1-209-277-2738	9600
1:207/705	The Halls of Valhalla BBS	Hemet CA	1-714-767-2442	9600
1:215/40	Moe	San Leandro CA	1-510-895-2843	9600
1:215/357	Combat Arms BBS	Castro Valley CA	1-510-537-1777	9600
1:216/21	House of Ill Compute	Boulder Creek CA	1-408-338-6860	9600
1:108/89	KIC	Cincinnati OH	1-513-762-1115	9600
1:108/145	Basselope West	Cincinnati OH	1-513-860-2277	9600
1:108/240	Listening Post	Cincinnati OH	1-513-474-3719	2400
1:110/10	Decker's Board	Dayton OH	1-513-439-9217	9600
1:110/35	Genealogy Ohio	Dayton OH	1-513-436-0400	9600
1:115/551	The Emporium System	Carpentersville IL	1-708-551-9275	9600
1:120/142	The Ultimate Force	Taylor MI	1-313-292-6167	9600
1:121/13	Computer Mania	Madison WI	1-608-276-7927	9600
1:121/99	NineJackNine	Madison WI	1-608-256-5697	9600
1:139/680	Foxy's Place	Appleton WI	1-414-739-8226	9600
1:154/222	Joe's Garage	Milwaukee WI	1-414-453-5145	9600
1:154/543	The Data Cache	Greenfield WI	1-414-543-9060	2400
1:154/321	The Edit Suite	Milwaukee WI	1-414-466-9983	9600
1:154/414	Radio Free	Milwaukee WI	1-414-352-6176	9600
1:157/2	Nerd's Nook II	Rocky River OH	1-216-356-1772	9600
1:157/607	The Radio Room BBS	Stow OH	1-216-686-8800	9600

Node	Name	Location	Phone	Speed
1:226/40	The CD-ROM BaseMent BBS	New Albany OH	1-614-855-3284	9600
1:226/60	Utilities Exchange BBS	Columbus OH	1-614-442-6695	9600
1:226/330	South Parking Lot	Columbus OH	1-614-351-2274	9600
1:231/30	The SouthSide BBS	New Whiteland IN	1-317-535-9097	9600
1:231/180	My Kinda Grouch BBS	Elwood IN	1-317-552-3397	9600
1:231/480	MacConnections	Indianapolis IN	1-317-290-1762	9600
1:234/2	Toledo's TBBS	Toledo OH	1-419-475-6003	9600
1:234/16	The Black Hole	Lima OH	1-419-222-6676	9600
1:236/13	DARK SHADOWS BBS	Angola IN	1-219-665-8767	9600
1:2200/141	Builders Workshop	Ypsilanti MI	1-313-483-1359	9600
1:2260/1	ORN Kentucky Hub	Worthington KY	1-606-836-1267	9600
1:2290/70	The Outer-Limits	Benton IL	1-618-439-9629	9600
1:2330/2007	The Coffee Break BBS	Spring Arbor MI	1-517-750-1847	2400
1:2370/1	Adventureland	Lexington KY	1-606-271-0558	9600
1:2370/10	PROF-BBS	Lexington KY	1-606-269-1565	9600
1:2370/12	OS/2 Connection	Lexington KY	1-606-223-7515	9600
1:163/239	Phoenix	Russell ON	1-613-445-3841	9600
1:163/215	BitByters	Rockland ON	1-613-446-6234	9600
1:163/506	R&D BBS	Hull PQ	1-819-772-2952	9600
1:167/134	Babillard Radio Amateur	Montreal PQ	1-514-728-1247	9600
1:167/230	VE2MMM Amateur Radio BBS	Pierrefonds PQ	1-514-624-5651	2400
1:167/281	7th ILLusion	Montreal PQ	1-514-338-1193	9600
1:167/116	Arcane BBS	Laval PQ	1-514-687-9586	9600
1:221/177	K-W Amateur Radio	Kitchener ON	1-519-578-9314	9600
1:221/275	Five Guys	Stratford ON	1-519-273-7668	9600
1:229/116	The Durham Board BBS	Whitby ON	1-416-666-4896	9600
1:229/412	The Assembly Line	Bowmanville ON	1-416-433-8923	9600
1:247/117	Air Waves	St. Catharines ON	1-416-984-4076	9600
1:248/301	Maxwell's Argentum Hammer	Brockville ON	1-613-345-0486	9600
1:250/202	The Beladau BBS	Toronto ON	1-416-975-1813	9600
1:250/401	MetroHUB 4	Rexdale ON	1-416-743-6703	9600
1:252/1	C.A.R.E. #1	Wasaga Beach	1-705-429-6036	9600
1:13/75	Al's Cabin	Milford PA	1-717-686-3037	9600
1:107/3000	HUB 300 EchoMail Coord	Central NJ	1-908-463-0315	9600
1:107/531	The Eagle's Nest	Lyndhurst NJ	1-201-939-2695	2400
1:109/118	NOVAC RBBS	Springfield VA	1-703-256-4777	2400
1:109/120	TIDMADT	Alexandria VA	1-703-370-7054	9600
1:109/104	ShanErin	Alexandria VA	1-703-941-8291	9600
1:109/229	Sentry Net BBS	Centreville VA	1-703-815-3244	9600
1:109/418	3 WINKs BBS	Gaithersburg MD	1-301-670-9621	9600
1:109/503	Foundation	College Park MD	1-301-935-4868	9600
1:129/89	BlinkLink	Pittsburgh PA	1-412-766-0732	9600
1:129/142	The Shadow Zone	Pittsburgh PA	1-412-231-7578	2400
1:150/220	WB3IKP BBS	Claymont DE	1-302-798-8186	9600
1:260/125	Block's BBS	Buffalo NY	1-716-832-9226	9600
1:260/126	Block's BBS	Buffalo NY	1-716-835-1621	9600

Node	Name	Location	Phone	Speed
1:260/132	The ERIDANUS BBS	N Tonawanda NY	1-716-694-9495	2400
1:260/224	The Silicon Metropolis	Farmington NY	1-716-398-3118	9600
1:260/232	Shack2	Rochester NY	1-716-288-5848	9600
1:260/233	Rat's Edge	Hamlin NY	1-716-964-7968	9600
1:260/312	OCC Micro/Tech BBS	Syracuse NY	1-315-492-6672	9600
1:260/328	Galaxia!	Phoenix NY	1-315-695-4436	9600
1:260/485	Dreamline	Binghamton NY	1-607-797-7508	2400
1:264/166	B&C BBS	Richmond VA	1-804-261-1819	9600
1:264/185	The Modem Medium BBS	Richmond VA	1-804-740-2665	2400
1:266/28	MoDem Corner	Edgewater Park NJ	1-609-877-0836	9600
1:266/32	Pinelands RBBS	Vincentown NJ	1-609-859-1910	9600
1:267/14	The Engineer's Studio	Saratoga Springs NY	1-518-587-9594	9600
1:267/54	The Final Frontier	Glens Falls NY	1-518-761-0869	9600
1:267/103	Radio FREQS'	Schenectady NY	1-518-377-7127	2400
1:267/132	Pabulum BBS	Broadalbin NY	1-518-883-4175	2400
1:268/110	Pompeiis BBS	Pen Argyl PA	1-215-863-7242	2400
1:268/202	NePa BBS	Berwick PA	1-717-759-1693	9600
1:270/101	The Other BBS	Harrisburg PA	1-717-657-7097	9600
1:270/311	P/T BBS	Cleona PA	1-717-272-6935	9600
1:270/911	Emergency Services BBS	Harrisburg PA	1-717-566-3500	9600
1:271/220	Brokedown Palace	Newport News VA	1-804-591-8537	9600
1:271/263	The Listening Post	Hampton VA	1-804-851-0616	2400
1:272/16	The WECA BBS WB2ZII	Pelham NY	1-914-738-6857	9600
1:272/39	Joe Brown's BBS	Mount Vernon NY	1-914-667-9385	9600
1:272/55	DataShack BBS	Eastchester NY	1-914-961-8959	9600
1:272/31	Red Onion Express	Wawayanda NY	1-914-342-4585	9600
1:273/215	Sophisticated Software BBS	Newtown PA	1-215-968-4998	9600
1:273/408	Airpower RYBBS	Lansdowne PA	1-215-259-2198	2400
1:273/208	The Datamax	Warminster PA	1-215-322-9193	9600
1:273/203	Satalink	Huntingdon Valley PA	1-215-953-9946	9600
1:273/714	System-2 BBS	Norristown PA	1-215-631-0685	9600
1:273/715	Alternative Lifestyles	Wyndmoor PA	1-215-242-4485	9600
1:273/709	Mystic's Mountain Maximus	Elkins Park PA	1-215-884-7449	9600
1:273/907	Tower BBS	Philadelphia PA	1-215-535-5917	9600
1:274/13	The Thunderbolt BBS	Falmouth Va	1-703-373-9289	9600
1:275/429	HandiNet B B S	Virginia Beach VA	1-804-496-3320	9600
1:275/99	The Apex	Chesapeake VA	1-804-436-3125	9600
1:275/17	The Computer Forum	Virginia Beach VA	1-804-471-0736	9600
1:278/702	NYC Fire Dept.	New York NY	1-212-964-8090	9600
1:278/712	Communication Specialties	New York NY	1-212-645-8673	9600
1:2600/120	MoROn MaNoR	Dover DE	1-302-735-8596	9600
1:2600/140	Theorem Beach	Viola DE	1-302-284-3570	9600
1:2601/100	W3NU Online	Sharon PA	1-412-346-5535	9600

Node	BBS Name	Location	Phone	Speed
1:2601/507	Mabel's Mansion	Sharon PA	1-412-981-3151	9600
1:2603/101	Brooklyn Perverts BBS	Brooklyn NY	1-718-853-8957	9600
1:2603/303	The Black Box II BBS	Staten Island NY	1-718-966-7651	9600
1:2604/110	The Nut House BBS	Ridgewood NJ	1-201-612-8594	9600
1:2604/201	Over The Edge	Montvale NJ	1-201-573-0719	9600
1:2605/123	Kin Ships	East Orange NJ	1-201-676-7066	2400
1:2606/406	Back Lounge of the Tour Bus	Hackettstown NJ	1-908-637-6336	9600
1:2612/107	Shuttle BBS	Charlotte Hall MD	1-301-884-0155	9600
1:280/3	ANARC BBS	Leawood KS	1-913-345-1978	2400
1:280/9	South Of The River BBS	Overland Park KS	1-913-642-7907	9600
1:280/12	KCATVG Support BBS	Kansas City MO	1-816-459-9752	9600
1:280/25	Howards Notebook	Belton MO	1-816-331-5868	9600
1:280/312	3-Times-7 BBS	Overland Park KS	1-913-599-6211	9600
1:280/316	The File Shop BBS	Kansas City MO	1-816-587-9936	9600
1:282/60	The Enterprise Board	Fridley MN	1-612-571-6280	9600
1:282/115	The Warehouse BBS	Minneapolis MN	1-612-379-8376	9600
1:282/62	Spare Computer	Minneapolis MN	1-612-824-2160	9600
1:282/100	HAM>link< RBBS	St Paul MN	1-612-426-0000	9600
1:283/610	Tri-State Data Exchange	Dubuque IA	1-319-556-4536	9600
1:283/125	Files R Us	Cedar Rapids IA	1-319-377-9257	9600
1:283/135	Cedar Valley DataNet	Cedar Rapids IA	1-319-393-4588	9600
1:285/21	ShadowFire Amiga	Omaha NE	1-402-734-1476	9600
1:288/6	TEXT bbs	Fargo ND	1-701-239-6048	9600
1:290/10	N0PBS Little System	Des Moines IA	1-515-265-0164	2400
1:291/6	Wichita BBS	Wichita KS	1-316-943-6030	9600
1:291/13	Q Continuum	Wichita KS	1-316-721-8466	9600
1:293/644	The Time Portal	RAPID CITY SD	1-605-348-4113	9600
1:294/1	Cmos Opus BBS	St Joseph MO	1-816-233-1357	9600
1:295/3	The Boarding House	Salina KS	1-913-827-0744	9600
1:104/114	The Dinosaur Board	Niwot CO	1-303-652-3595	9600
1:104/115	King's Market BBS	Louisville CO	1-303-665-6091	9600
1:104/28	Pinecliffe HST DS	Boulder CO	1-303-642-0703	9600
1:104/325	Master Control (NSN)	Avon CO	1-303-949-3253	9600
1:104/424	Ready Room: The Next Generation	Denver CO	1-303-755-1681	9600
1:104/477	Jaguar's Networking Labs	Denver CO	1-303-377-2371	9600
1:104/512	The Peacock's Nest	Aurora Co	1-303-680-0509	9600
1:104/666	The Comm-Post	Denver CO	1-303-534-4311	9600
1:104/810	Electronic Library	Denver CO	1-303-935-6323	9600
1:104/914	The Whistlestop	Denver CO	1-303-592-8380	2400
1:114/36	Nighthawk BBS	Phoenix AZ	1-602-582-1127	9600
1:114/148	T.V. BBS	Glendale AZ	1-602-930-8542	9600
1:114/22	Rare Readers BBS	Tempe AZ	1-602-756-2855	9600
1:114/52	Steve's One Stop DLG Shop	Phoenix AZ	1-602-788-7144	9600
1:128/61	Ham Radio BBS	Colorado Springs CO	1-719-390-5318	2400
1:101/99	Real Times #1	Boston MA	1-617-783-8820	9600
1:101/275	DX Online	Lynn MA	1-617-592-8404	2400
1:101/460	VI/BUG	Holbrook MA	1-617-767-2909	9600
1:101/470	Tom's BBS	Milton MA	1-617-698-8734	9600

Node	BBS Name	Location	Phone	Speed
1:132/300	OCI Online Communications Inc	Bangor ME	1-207-990-3511	9600
1:141/455	The Planet Earth	Bridgeport CT	1-203-335-7742	9600
1:141/485	The Soft Parade	Shelton CT	1-203-924-5603	9600
1:141/730	Treasure Island	Danbury CT	1-203-791-8532	9600
1:141/725	Source of Magic	Ridgefield CT	1-203-431-4687	9600
1:142/885	The Water Hole BBS	Ellington CT	1-203-875-7071	9600
1:142/911	Hart-Metro Fido	Wethersfield CT	1-203-563-6455	9600
1:321/109	Pioneer Valley PCUG1	Amherst MA	1-413-256-1037	9600
1:321/155	Recipe Corner BBS	Greenfield MA	1-413-774-3601	2400
1:321/307	The Macintosh Only BBS!	Springfield MA	1-413-532-1387	9600
1:321/203	VETLink#1	Pittsfield MA	1-413-443-6313	9600
1:321/214	Berkshire Hills BBS	Adams MA	1-413-743-1111	9600
1:322/14	WayStar	Marlborough MA	1-508-481-7147	9600
1:323/114	Road Runner BBS	West Warwick RI	1-401-821-1457	9600
1:323/115	ImageNet BBS	Coventry RI	1-401-822-3060	9600
1:324/127	Computer Castle	Haverhill MA	1-508-521-6941	9600
1:324/130	The Wizards Lair	Merrimac MA	1-508-346-8213	9600
1:324/175	Lost In The Supermarket	Peabody MA	1-508-531-8416	9600
1:324/278	Daves Opus	Lowell MA	1-508-454-3864	9600
1:324/290	Ken's BBS	Dracut MA	1-508-957-5408	2400
1:328/104	The Lobster Buoy	Bangor ME	1-207-945-9346	9600
1:105/380	The Digital Amateur	Gaston Or	1-503-359-5111	2400
1:105/405	Et Cetera	Gresham OR	1-503-663-1459	9600
1:105/642	Darkstar System	Longview WA	1-206-578-1157	9600
1:105/7	The ROSE	Portland OR	1-503-286-3855	9600
1:105/52	The Garden Pond	Portland OR	1-503-735-3074	9600
1:105/87	Sherri L Knobel Memorial	Portland OR	1-503-244-5711	9600
1:105/97	Dialogues BBS	Tigard OR	1-503-244-0977	2400
1:342/55	Probable Fate Systems	Edmonton Alta	1-403-453-2223	2400
1:138/102	Awakening	Tacoma WA	1-206-582-5579	9600
1:343/5	Brier Opus	Brier WA	1-206-743-9452	2400
1:343/40	Top Hat	Seattle WA	1-206-244-9661	9600
1:343/58	Silver Lake	Everett WA	1-206-338-3723	9600
1:343/47	The Boardwalk	Auburn WA	1-206-941-4531	9600
1:344/17	Electronic Library	Wenatchee WA	1-509-663-5232	9600
1:346/3	Radio Therapy BBS	Spokane Wa	1-509-534-7924	9600
1:346/8	The Think Tank II	Spokane WA	1-509-244-6446	9600
1:346/73	RF-X	Spokane Wa	1-509-325-0270	2400
1:140/1	Net Echo Coordinator	Saskatchewan	1-306-585-1958	9600
1:153/9	Basic'ly BBS	Surrey BC	1-604-589-8561	9600
1:153/715	The BandMaster	Vancouver BC	1-604-266-7754	9600
1:153/726	PCS BBS	North Vancouver BC	1-604-671-3028	9600
1:153/739	The Ham Shack	Burnaby BC	1-604-298-9225	2400
1:153/792	2ND DATA CIRCUIT	Vancouver BC	1-604-984-6036	9600
1:352/458	The Elders' Council BBS	Tumwater WA	1-206-357-8992	2400
1:352/777	The ICDMnet IHQ	Olympia WA	1-206-866-3621	9600
1:18/230	Ancestry TBBS	Sebring FL	1-813-471-0552	9600
1:112/25	SoftWare Exchange	Jacksonville FL	1-904-389-8212	9600
1:116/1	Transfer Station	Nashville TN	1-615-297-5611	9600

Node	Name	Location	Phone	Speed
1:116/3000	The Homestead	Nashville TN	1-615-385-9421	9600
1:123/13	The NiteMare BBS	Memphis TN	1-901-754-9823	9600
1.123/19	Memphis Mail Hub	Memphis TN	1-901-353-2429	9600
1:133/108	Aviation OnLine	Atlanta GA	1-404-740-9336	9600
1:133/608	Garden of Eden	Atlanta GA	1-404-256-0204	9600
1:133/804	Midnight Madness	Peachtree City GA	1-404-487-5329	2400
1:135/23	Telcom Central	Miami Lakes FL	1-305-828-7909	9600
1:135/58	Weatherman	Miami FL	1-305-254-8344	9600
1:151/102	Micro Message Service	Raleigh NC	1-919-772-7654	9600
1:151/124	CAROLINA TRACON	Cary NC	1-919-469-1864	9600
1:151/602	Dark Star	Winston-Salem NC	1-919-766-1072	9600
1:360/1	Augusta Forum	North Augusta SC	1-803-279-4124	9600
1:362/411	Pelican Key!	Chattanooga TN	1-615-877-0411	2400
1:363/29	Gourmet Delight	Orlando FL	1-407-649-4136	9600
1:363/18	Cornucopia TBBS	Winter Park FL	1-407-645-4929	9600
1:363/123	The Listening Post	Oviedo FL	1-407-365-9809	9600
1:369/8	The Catwalk BBS	Davie FL	1-305-370-3528	9600
1:373/2	Gateway	Huntsville AL	1-205-880-7723	9600
1:374/51	GadgetTech BBS	Vero Bch FL	1-407-562-0580	2400
1:374/1	TechTalk	Titusville FL	1-407-269-5188	9600
1:374/73	The Bear's Cave	Titusville FL	1-407-383-9372	9600
1:375/33	The Perfect Blend	Montgomery AL	1-205-244-0254	9600
1:376/24	Fort Mill	Fort Mill SC	1-803-548-0900	9600
1:377/817	The Phoenix BBS	Tampa Fl	1-813-840-9101	9600
1:379/1	Transporter Room Pod II NEC379	Charlotte NC	1-704-567-9594	9600
1:379/10	Carolina Forum	Charlotte NC	1-704-563-5857	9600
1:379/16	AET's BBS	Mint Hill NC	1-704-545-7076	9600
1:379/37	Borderline! BBS	Kannapolis NC	1-704-938-6207	2400
1:379/1202	Superstar Connection	Albemarle NC	1-704-982-9296	2400
1:3603/20	Mercury Opus NEC3603	St Petersburg FL	1-813-321-0734	9600
1:3603/326	SPPE	St Petersburg FL	1-813-525-2326	9600
1:3613/10	BackWoods BBS	Upatoi Ga	1-706-561-6106	9600
1:3615/24	Poor Valley	Knoxville TN	1-615-573-1569	9600
1:3619/15	The Modem Zone BBS	New Port Richey FL	1-813-372-1146	2400
1:3619/21	Inner Sanctum	New Port Richey FL	1-813-849-2998	9600
1:3625/465	The Data Connection!	Mobile AL	1-205-602-0917	9600
1:3628/5	Wilmington*80	Wilmington NC	1-919-763-1850	2400
1:3641/1	Durham Net NEC	Durham NC	1-919-286-7738	9600
1:106/18	The ComPort BBS	Houston TX	1-713-947-9866	9600
1:106/99	Cloud Nine BBS	Houston TX	1-713-855-4385	9600
1:106/357	Conch Opus	Houston TX	1-713-667-7213	9600
1:106/2000	COMM Port One	Sugar Land TX	1-713-980-9671	9600
1:106/3198	CSI Online	Conroe TX	1-409-321-3198	9600
1:106/8324	Far Point Relay Ham BBS	Katy TX	1-713-463-8324	9600
1:106/8325	Far Point Back Door	Katy TX	1-713-463-4434	9600
1:106/9788	Ye Olde Inn III	Alvin TX	1-713-331-3056	9600
1:124/3106	Squirrel Talk	Irving TX	1-214-594-7911	9600
1:124/5118	Smoked Armadillos	Richardson TX	1-214-669-9645	9600
1:124/7009	Comm Port One	Garland TX	1-214-226-1181	9600
1:124/7012	Terminator	The Colony TX	1-214-625-2448	9600
1:130/22	The ARChive	Burleson TX	1-817-447-1969	9600

Node	Name	Location	Phone	Speed
1:147/2001	Retriever's Retreat	Edmond OK	1-405-359-1540	9600
1:147/31	The General Store	Oklahoma City OK	1-405-943-8638	9600
1:160/1	The SeaHorse/2	Corpus Christi TX	1-512-994-9643	9600
1:160/80	Runtime Error	Portland TX	1-512-643-7775	9600
1:170/405	Cat-Tastrophe Corner	Tulsa OK	1-918-663-1249	9600
1:380/5	My Secret Garden RBBS	Shreveport La	1-318-865-4503	9600
1:382/1	Crystal Palace	Lake Travis TX	1-512-335-7949	9600
1:382/40	The Antenna Farm	Austin TX	1-512-444-1052	9600
1:386/1	Beacon Terra 1	Galveston TX	1-409-765-6632	9600
1:387/505	General Store	San Antonio TX	1-512-520-4525	9600
1:388/8	Phantom Zone	Lorena TX	1-817-857-3523	9600
1:388/20	The Mule Barn	Waco TX	1-817-756-7565	9600
1:388/21	Hello World	McGregor TX	1-817-840-2140	9600
1:390/22	Miller's Crossing BBS	Slidell LA	1-504-649-7388	9600
1:390/5	WSTPC	Sun LA	1-504-886-2157	9600
1:392/6	The CD ROM-BBS	Abilene Tx	1-915-673-8014	9600
1:396/15	The Silver Streak RBBS	New Orleans LA	1-504-888-6515	9600
1:396/19	Tri-Parish Exchange	LaPlace LA	1-504-652-7014	9600
1:396/22	The Bowling Alley RBBS	New Orleans LA	1-504-466-0908	9600
1:396/28	La. Medsig	Harahan LA	1-504-738-3900	9600
1:396/47	The Big Easy	New Orleans LA	1-504-464-0289	2400
1:396/65	The Digital Cottage	New Orleans LA	1-504-897-6614	2400
1:3800/18	The Chatter Box	Baker LA	1-504-775-7825	9600
2:200/110	SweDX	Malmoe	46-40-973280	9600
2:201/111	Capital City	Haninge	46-8-7411244	9600
2:201/112	CetuSoft	Stockholm	46-8-7360455	9600
2:201/119	GET	Lidingo	46-8-7317355	9600
2:241/5306	zaphods bbs i	Bonn	49-228-262894	9600
2:241/5603	Funboard	Velbert	49-2051-81781	9600
2:241/7200	Bodensee Connection	Friedrichshafen	49-7541-53132	9600
2:243/50	Higli-Box	Koblenz	49-261-69205	9600
2:246/2	C.A.C.-BOX Muenchen-QuickBBS Help-Node GER	Muenchen	49-89-7469379	9600
2:249/29	Guenter's Mailbox	Mainz-Kostheim	49-6134-24816	9600
2:250/116	LOOK NORTHWEST	Nelson	44-282-698380	9600
2:252/129	MetNet Triangle	Hull UK	44-482-473871	2400
2:252/108	Jolly Roger BBS [1]	London UK	44-81-742-1640	9600
2:252/136	Jolly Roger BBS [2]	London UK	44-81-995-5829	9600
2:252/21	(-: Golly! :-)	Twyford	44-734-320812	9600
2:253/157	TUG II	Droitwich UK	44-905-775191	9600
2:255/402	SYSTEM X	York UK	44-904-612934	9600
2:259/2	Alba Maximus	Barrhead UK	44-41-880-7845	9600
2:440/32	BoysVille BB	Leicester UK	44-533-559171	9600
2:263/151	TOPPSI	Dublin Ireland	353-1-711047	9600
2:281/527	Contrast	Den Haag NL	31-70-3234903	9600
2:284/207	Filelift	Eindhoven NL	31-40-123677	9600
2:291/705	PMS BBS	Brussels B	32-2-3758656	9600
3:632/309	Melbourne PC User Group	Sth Melbourne VIC	61-3-510-6180	9600
3:634/393	Spectrum Radio	Hawthorn Victoria	61-3-819-9167	2400
3:681/854	The Phone Box #1	Inglewood	61-8-380-5505	9600

3:681/855	The Phone Box #2	Inglewood	61-8-380-5606	9600
3:681/864	The North Star	Elizabeth Park	61-8-252-1082	2400
3:681/869	Adelaide Mailbase	Walkerville	61-8-269-1242	9600
3:711/907	NSS Sydney Australia	Waverton NSW	61-2-954-0934	9600
3:711/430	Coastal Opus	Springfield NSW	61-43-23-2275	9600
3:712/619	Cross Facts BBS	Abbotsford NSW	61-2-712-3910	9600
3:713/604	Food For Thought	Parramatta NSW	61-2-683-6093	9600
3:713/605	Shortwave Possums	Dural NSW	61-2-651-3055	2400
3:713/635	Vulcan's World BBS	Westmead NSW	61-2-635-1204	9600
3:714/207	Palantir	Avalon NSW	61-2-975-3355	9600
3:640/206	LANDS MultiLine	Woolloongabba QLD Australia	61-7-393-0311	9600

CHAPTER 11
SHORT WAVE EQUIPMENT SOURCES

The following companies are known to cater for the shortwave broadcast listener market and all carry more than one brand of shortwave receiver or antenna. We have spoken with each of the companies listed below or the dealer has been suggested by WRTH readers during 1992. Inclusion of any company does NOT imply an automatic recommendation by the World Radio TV Handbook, nor does exclusion imply that the dealer did not qualify for mentioning. Regular advertisers in the WRTH are shown in bold type. Addresses are included for information only. Further suggestions from readers would be welcome. Write to: Equipment Survey, WRTH Editorial Office, P.O. Box 9027, 1006 AA Amsterdam.

Note that any phone numbers given begin with the country code for that country, e.g. if dialling a UK number from within Britain change the "44" into a "0".

(IRC = International Reply Coupon, available from most larger post offices.)

Amalgamated Wireless, P.O. Box 830, Wellington, NEW ZEALAND.

Andy's Funkladen, Admiralstrasse 119, D-2800 Bremen 1, GERMANY. Tel: +49 421 353060. Fax: +49 421 372714.

ARE Communications, 6 Royal Parade, Hanger Lane, London W5A 1ET. Tel: +44 81 997 4476. Fax: +44 81 991 2565. Sell Icom, Yaesu, Kenwood, Sony, JRC, Drake.

Atlantic Ham Radio Ltd, 368 Wilson Avenue, Downsview, Ontario, Canada M3H 1S9, CANADA. Tel: +1 416 636 3636. Fax: +1 416 631 0747.

Barry Electronics Corp, 512 Broadway, New York, NY 10012, USA. Tel: +1 212 925-7000. Fax: +1 212 925 7001

Bogerfunk, Grundesch 15, D-7960 Aulendorf/Steinenbach, GERMANY. Tel: +49 7525 451. Fax: +49 7525 2382

Bredhurst Electronics Ltd, High Street, Handcross, West Sussex, RH17 6BW, ENGLAND. Tel: +44 444 400786. Fax: +44 444 400604. Mainly sells Lowe, Icom.

Campbell Appliances, 116 Picton Street, Howick, Auckland, NEW ZEALAND. Tel: +64 9 5349159.

Century 21 Communications Inc, 4610 Dufferin Street, Unit 20-B, Downsview, Ontario M3H 5S4, CANADA.

Com-Centre, Shop 1 & 2, 275 New Windsor Rd, Blockhouse Bay, Auckland 7, NEW ZEALAND. Tel: +64 9 873 213.

Com-West Radio Systems Ltd, 8179 Main Street, Vancouver, BC, V5X 3L2, CANADA. Tel: +1 604 321-1833. Fax: +1 604 321 1833. Sells Icom, Kenwood, Yaesu, Ilenco, Ham radio equipment.

Doeven Electronics, Schutstraat 58, 7901 EE Hoogeveen, THE NETHERLANDS. Tel: +31 5280 69679.

Dubberley's on Davie, Ltd, 920 Davie St, Vancouver, BC V6Z 1B8 CANADA. Tel: +1 604 684-5981. Fax: +1 604 684 7520.

Duty Free Electronics Ltd, Amsterdam (Schiphol) Airport Duty-Free Shopping Centre. Tel: +31 20 6540700.

Electronic Center, Inc, Ross at Central Expressway, Dallas, TX 75201, USA. Tel: 1 800 441-0145

Electronic Equipment Bank, 323 Mill Street N.E., Vienna, VA 22180 USA. Tel: +1 703 938 3350 or toll-free (in US only) 1 800 368 3270.

Emtronics, 92-94 Wentworth Avenue, Sydney 2000, N.S.W. or 288-294 Queens Street, Melbourne 3000 VIC, AUSTRALIA. Tel: +61 3 6708551.

Fritzel Kurt, Siemenstrasse 2, D-6708 Neuhofen/Pfalz, GERMANY. Tel: +49 62 3652044. Fax: +49 62 3652236. Antenna specialist.

Galaxy Electronics, Box 1202, 67 Eber Avenue, Akron, OH 44309, USA. Tel: +1 216 376 2402. Sells Icom, Kenwood, JRC.

Gilfer Associates, 52 Park Avenue, P.O. Box 239, Park Ridge, NJ 07656, USA. Tel: +1 201 391 7887 or in USA only 1 800 445 3371. Full range of equipment + own publications.

Grove Enterprises, 140 Dog Branch Road, Brasstown, NC 28902, USA. Tel: +1 704 837 9200 or (in USA only) 1 800 438 8155.

Ham Radio Outlet, 933 North Euclid, Anaheim, CA 92801, USA. Tel: 1 714 533 7373 Fax: +1 714 533 9485. 11 branches across US. Sells Icom, Kenwood, Yaesu, Sangean, Sony.

Hamrad Pty Ltd, 114 Buitengragt, Capetown 8001, REPUBLIC OF SOUTH AFRICA. Tel: +27 21 24 7060. Yaesu agent

Katuka Musen, 3-13-7, Sotokanda, Chiyoda-ku, Tokyo 101, JAPAN. Tel: +81 3 255 5461. All Japanese Equipment.

KW Communications, Chatham Road, Sandling, Kent ME14 3AY, ENGLAND. Tel: +44 622 692773. Fax: +44 622 764614.

Lieberman Electronics (Pty) Ltd, Malotune, # 5th Street, Thora Crescent, Wijnberg, 2090 (Postal address: P.O. Box 707, Bergvlei 2012) REPUBLIC OF SOUTH AFRICA. Tel: +27 11 8876580. Agents for Kenwood & Philips.

Lowe Electronics Ltd., Bentley Bridge, Chesterfield Rd, Matlock Derbyshire, DE4 5LE, ENGLAND. Tel: +44 629 580800. Agents for Kenwood, JRC.

Mach Sound Company, 68 Orchard Rd Number 03-31, Plaza Singapura 0923, SINGAPORE. Tel: +65 336 1710.

Mil-Spec Communications, P.O. Box 461, Wakefield, RI 02880, USA. Tel: +1 401 783 7106. Sells Sony, Panasonic, Kenwood, Collins, Uniden, Radio Shack.

Nevada Communications, 189 London Road, Portsmouth, PO2 9AE, ENGLAND. Tel: +44 705 662145. Fax: +44 705 690626.

Norham Radio, 4373 Steeles Ave W, North York, Ontario, CANADA M9L 2W1. Tel: +1 416 667 1000. Fax: +1 416 667 9995.

Pacific Radio Pte Ltd, 37 Jalun Besar, Singapore 0820, SINGAPORE. Tel: +65 294 0852.

Peter Lai Pte Ltd, 304 Orchard Rd, Number 02-09 Lucky Plaza, Singapore 0923 SINGAPORE. Tel: +65 235 2073.

Radio Communications Center, Amsterdamsestraatweg 561-563, 3553 EG Utrecht, THE NETHERLANDS. Tel: +31 30 433835.

Radio Shack Ltd, 188 Broadhurst Gardens, London NW6 3AY, ENGLAND. Tel: +44 71 624 7174. Sells Icom, Drake.

Radio West, 850 Anns Way Drive, Vista, CA 92083 USA. Tel: +1 619 726 3910.

Rico Pte Ltd, 80 Genting Lane Number 10-10, Genting Block, Singapore 1334, SINGAPORE. Tel: +65 745 8472.

Raycom Communications Systems Ltd, International House 963, Wolverhampton Road, Oldbury, West Midlands, B69 4RJ, ENGLAND. Tel: +44 21 544 6767. Fax: +44 21 544 7124. Make modifications to Yaesu FRG-9600 and Kenwood RZ-1.

Shortwave Centre, 95 Colindeep Lane, Sprowston, Norwich, NR7 8EQ, ENGLAND. Tel: +44 603 788281. Fax: +44 603 788281.

Solar Light Company, 6655 Lawnton Avenue, Philadelphia, PA 19126 USA. Tel: +1 215 548 4747 or +1 215 927 4206. Offers the CR-2020 Timer Cassette Controller. This attaches simply to the Sony ICF-2001, ICF-2001D, ICF-7600 (all types except ICF-7600A), ICF-2003, Panasonic RF-B60, or Sangean ATS-803(A). The controller allows the built-in standby function to control an external cassette tape recorder. Price is US$37.95 in USA including shipping. Outside USA, apply in advance for price details.

South Midlands Communications Ltd, SMC HQ, School Close, Chandlers Ford Industrial Estate, Eastleigh, Hants S05 3BY, ENGLAND. Tel: +44 703 255111. Fax: +44 703 263507.

Spectronics, Inc, 1009 Garfield St., Oak Park, IL 60304, USA. Tel: +1 312 848-6777.

Strand Audio, 33-39 Talavera Road, North Ryde, Sydney, NSW 2113, AUSTRALIA. Tel: +61 2 887 6666.

Surrey Electronics, The Forge, Lucks Green, Cranleigh, Surrey GU6 7BG, ENGLAND. Tel: +44 483 275997. Fax: +44 483 276477. Manufacturers of special SW synchronous detector boards, and modifications to FRG-8800.

SW Horizons, #61, 52152 Range Road 210, Sherwood Park, Alberta T8G 1A5, CANADA. Tel: +1 403 922 2872.

Tedelex Sound & Vision, P.O. Box 10525, Johannesburg 2000, SOUTH AFRICA. Tel: +27 11 683 5800. Sony agent.

Teletech Services, #05-06 GRTH Building, 66-68 East Coast Road, Singapore 1542, SINGAPORE.

Tomlihan Pte Ltd, Siong Huat Building, 240 McPherson #06-03, Singapore 1334, SINGAPORE. Mainly Kenwood.

TVA Televisioapu Oy, Helsingika 30, SF-00530 Helsinki, FINLAND.

Universal Shortwave Radio, 6380 Americana Parkway, Reynoldsburg, OH 43068, USA. Tel: +1 614 866 4267 or in USA only 1 800 431 3939. Fax: +1 614 866 2339. Full range of equipment + own publications.

Waters & Stanton Electronics, 22 Main Road, Hockley, Essex, ENGLAND. Tel: +44 702 206835.

MILITARY SURPLUS RECEIVERS

These are recommended for a restricted group of listeners, preferably with a technical background. These sets were manufactured for military use and are up to 50 years old. They do not look attractive but performance can often be compared with the "B" and even "C" category of receiver. They were built to a specification and not to a price. The sets had type numbers such as National HRO 50, the BC 348, Murphy B40 (came in two versions), RCA AR88, and Collins R-390A. These receivers are getting very scarce, and will probably need a lot of work to restore them. A new wave of "surplus" sets has emerged recently, especially in Europe. Many sets purchased in the 1960's (e.g. Racal RA17) are now coming onto the market. When properly aligned, they provide good performance and thousands are still in regular use. They all work using valves (tubes) and replacements may be hard to find now, though they are still available through mail order houses in Europe or in radio shops in Asia, especially in the Indian subcontinent. Certain companies specialize in supplying old parts. Here are a few addresses.

AJH Electronics, 151a Bilton Road, Rugby, Warks, CV22 7AS, ENGLAND. Tel: +44 788 76473/71066. Surplus Communications Receivers.

Antique Radio Parts, Box 42, Rossville, IN 46065, USA. Antique parts and literature.

Colomor Electronics Ltd, 170 Goldhawk Road, London W12, ENGLAND. Fax: +44 81 749 3934. Source of valves and transistors.

Fair Radio Sales Co., P.O. Box 1105, 1016 East Eureka Street, Lima OH 45802, USA. Tel: +1 419 227 6573 or +1 419 223 2196. Fax: +1 419 227

1313. Excellent surplus catalogue produced once a year around March, with a supplement later in the year. Send return postage or call for details.

"Hollow State News". This Newsletter was revived in early 1991 and is one of the few sources of information on restoring military surplus receivers, such as the R390A. The Newsletter is quarterly for a nominal US$5.00 a year (in US), overseas rates on request. More information from Hollow State News, Ralph Sanserino, 11300 Magnolia #43, Riverside, CA 92505, USA

McHahon's Vintage Radio, Box 1331, North Highlands, CA 95660, USA. Tel: +1 916 332 8262. Has a list of numerous vintage radio publications and a circuit diagram finding service.

Modern Radio Labs, Box 1477, Garden Grove, CA 92642, USA. Supplier of parts and literature.

Old Time Radio Co., 2445 Lyttonsville Rd., Silver Spring, MD 20910, USA. Antique parts.

PM Components, Selectron House, Springhead Enterprise Park, Springhead Rd, Gravesend, Kent DA11 8HD, ENGLAND. Tel: +44 474 560521. Source of radio valves (tubes).

Steinmetz Ltd., 7519 Maplewood Ave, Hammond, IN 46324, USA. (catalogue 2 IRC's)

"The Vintage Wireless Company", Tudor House, Cossham Street, Mangotsfield, Bristol BS17 3EN, ENGLAND. Tel: +44 272 565472. The "Antique Wireless Newsheet" covers news in this field, and a 12 issue annual subscription costs £5.00 post paid in UK, £6.00 overseas airmail.

Wilson Valves, Peel Cottage, Lees Rd, Mossley, Ashton-under-Lyne, OL5 0PG, ENGLAND. Source of radio valves.

VINTAGE RADIO SOCIETIES

Some groups exist that concern themselves with old receivers, though not necessarily shortwave or military surplus sets. Here are some of the most active:

The Antique Wireless Association Inc. is one of the largest radio collecting organizations in the world. It boasts an excellent magazine "The Old Timer's Bulletin" as well as an annual flea market and seminar session. Subscriptions are US$10.00 per year in US, US$12.00 per year outside. Further details from AWA, Box "E", Breesport, New York, NY 14816, USA.

There is also another organization called the "Antique Radio Club of America". They also have a regular newsletter (The Gazette) and an annual convention. The address is ARCA, 81 Steeplechase Rd, Devon, PA 19333, USA. Enclose return postage and an self addressed envelope.

Historic Radio Society of Australia, c/o Rex Wales, 24 Park Lane, Mt. Waverley, VIC 3149, AUSTRALIA. Publishes a regular newsletter.

British Vintage Wireless Society, c/o Gerald Wells, 23 Rosendale Rd, Dulwich, London SE21 8DS, ENGLAND. Tel: +44 81 670 3667. Excellent quarterly Newsletter full of fascinating articles and information. Excellent museum in Dulwich, South London. Membership details for return postage, and self addressed envelope.

Nederlandse Vereniging voor de Historie van de Radio, c/o J.E.J.W. Hermans, Paulus Potterstraat 19, 6814 KT Arnhem, THE NETHERLANDS. Tel: +31 8376 13016. This Dutch society with some 1100 members publishes a DUTCH language bulletin 4 times a year. It contains a lot of fascinating articles. Details for return postage, but you'll need to understand Dutch to follow their work.

New Zealand Vintage Radio Society, c/o Bryan Marsh, 20 Rimu Road, Mangere Bridge, Auckland, NEW ZEALAND. Tel: +64 9 667-712. Quarterly bulletin dedicated to the preservation and restoration of early radio equipment. Subscriptions in New Zealand are NZ$15.00 per year. Overseas rates are extra depending on airmail postage.

OTHER SOURCES

Antique Electronic Supply produce an interesting catalogue of hard-to-get components for early radios (mainly North American types). They also carry a line of books on restoring radios. More information from Antique Electronic Supply, 6221 South Maple Ave, Tempe, AZ 85283, USA. Tel: +1 602 820 5411.

Chevet Books, 157 Dickson Rd, Blackpool FY1 2EU, Lancs, ENGLAND. Tel: +44 253 751858. This publisher makes a regular listing of manual and catalogue reprints, and will search for specific titles. Current catalogue also contains details of a large valve (tube) collection for sale.

McMahon's Vintage Radio, Box 1331, North Highlands, CA 95660, USA. Tel: +1 916 332 8262 has a list of numerous vintage radio publications and a circuit diagram finding service.

New Wireless Pioneers-Bampton Books, Box 398, Elma, New York, NY 14059, USA. Tel: +1 716 681 3186. This company publishes 2 catalogues a year. They contain 45+ pages consisting of old, rare and unusual books on early radio, wireless, electricity, telephony, telegraphy, vintage audio and television. Also included are magazines, catalogues and ephemera (sometimes autographs). These items are priced anywhere from US $2.00 to $500. They send a catalogue free on request and individual wants lists are encouraged..

The Vintage Wireless Co Ltd, Tudor House, Cossham Street, Mangotsfield, Bristol, BS17 3EN, ENGLAND. Tel: +44 272 565472. Fax: +44 272 575442. Largest UK stocklist of technical service information on vintage radios including communications, electronic valves, components etc. The "Antique Wireless Newsheet" covers news in this field, and a 12 issue subscription costs £5.00 post paid in UK, £6.00 overseas airmail. Free catalogue available, IRC's appreciated.

Vestal Press Ltd, P.O. Box 97, 320 N.Jensen Rd, Vestal, New York, NY 13850, USA. Tel: +1 607 797 4872. This is one of the few publishing houses that specializes in nostalgia, and radio books in particular. The group also produces cassettes of mechanical music. They produce an excellent bi-annual newsletter. One of the best Vestal projects currently under way is a series entitled Radio Manufacturers of the 1920's. Author Alan Douglas has put out two volumes so far, full of beautiful illustrations.

SERVICE SHEET SOURCES

If you plan to repair your radio receiver yourself then the following addresses may also be of help. However, we wish to warn you that unless you know precisely what you are doing, adjusting coils and transformers will usually result in a disaster. In cases where the receiver is connected to the mains electricity inexperienced tampering can be lethal. We cannot emphasize this point too strongly! Most people who have tried to save money, have ended up with a bigger bill later. On the other hand, if you want a service manual for your receiver, we suggest you write to the "Consumer Service Department" of the manufacturer that made your set. Many can and do supply manuals for a modest cost, up to about seven years after the set is discontinued.

For older and surplus sets (see above) the following might help. Please mention World Radio TV Handbook, and note that any phone numbers given begin with the international code for that country, e.g. if dialling a UK number from within Britain change "44" into "0".

The Technical Information Service, 76 Church Street, Larkhall, Lanarks ML9 1HE, SCOTLAND. Tel: +44 698 883334. Return postage and self-addressed envelope brings Newsletter. Manual costs £2.50 plus postage.

Hamilton Radio, 47 Bohemia Road, St Leonards, Sussex, ENGLAND. Service sheets from £0.50. Catalogue £0.25. We suggest readers outside UK enclose 2 IRC's for postage.

C.Caranna, 71 Beaufort Park, London NW11 6BX, ENGLAND. Tel: +44 81 458 4882. Stamped, addressed envelope for catalogue.

Pax Manufacturing Corp., 100 East Montauk Highway, Lindenhurst, NY 11757, USA. Specializes in Hammarlund line, also sells parts as well as manuals.

Mr.P.Wamock, 45 Rothwell, Ottawa K1J 7G7, CANADA. Catalogue for return postage.

Mr.M.Consalvo, 7218 Roanse Drive, Washington, D.C. 20021, USA. Collection of military surplus equipment. He will handle specific requests only. Send self-addressed envelope and return postage.

Bill Slep Electronics Co, P.O. Box 100, Otto, NC 28763, USA. Supplier of manuals and spare parts. Return postage for details.

Technical Manuals of US Army, Navy and Air Force equipment after WWII can be obtained from National Archives & Record Services, Washington,

D.C. 20408, USA. Photocopies cost 30 cents per page, with a minimum order of US$6.00 by mail.

Vintage Wireless Company, Tudor House, Cossham Street, Mangotsfield, Bristol, BS17 3EN, ENGLAND. Tel: +44 272 565472. All enquiries must enclose return postage. See also entry above.

FURTHER SOURCES OF RECEIVER INFORMATION

1. World Radio TV Handbook 1993. No.of pages: 608. Publisher: Billboard Publications Inc, 1515 Broadway New York NY 10036 USA. Price: $19.95 in the USA. Edited in Amsterdam, The Netherlands, this is THE reference source for the shortwave listener; a sort of telephone book of the airwaves. This year's edition is expanded to 608 pages containing frequencies, schedules, addresses, and background information on the business. Also features independent receiver and antenna tests, a look at propagation, clubs and broadcasting organizations. 1993 edition will be published in December 92. In USA: Watson-Guptill, 1515 Broadway, New York NY 10036 USA.

2. "Shortwave Receivers Past and Present" is an excellent low-cost reference book designed to assist the shortwave listener who is thinking of buying a used shortwave set. Key specifications (and usually a photo) are given for over 200 recently manufactured models. Some brief comments are also included. Some idea of the second-hand value is given, though this relates specifically to the United States. The edition was published in January 1988, so some of the second-hand price data is out of date. However it is still a very interesting read and costs US$8.95. Shipping costs are US$1.00 extra in the US, from other areas on application. Further information from the editor Fred Osterman at Universal SW Radio, 6830 Americana Parkway, Reynoldsburg, Ohio 43068 USA. Tel: +1 614 866 4267.

3. "A DXer's Technical Guide". Published in 1982, this 2nd edition is a 120 page collection of articles by various North American authors. Mainly aimed at the medium wave (AM) DXer, there are many points that would appeal to the technically minded shortwave listener too. Available from the author Nick Hall Patch, P.O. Box 21074, Seattle WA 98111, USA. Price is US$6.50, $9.50 elsewhere postpaid.

4. "Monitoring Times". Monthly newspaper-style magazine put out by mail-order company Grove Enterprises. Contains a lot of news about new equipment. For subscription details write to Monitoring Times, P.O. Box 98, Brasstown, NC 28902 USA.

CHAPTER 12
WRTH EQUIPMENT GUIDE READERS SURVEY

We want the WRTH to give you the best and most reliable information in the business. To maintain the standard we'd like your honest feedback to some simple questions. All complete entries will be eligible for a draw in which the prizes will be: The top prize is a digital portable receiver to the value of US$300, or cash equivalent. There will be three runner-up prizes of a 1994 World Radio TV Handbook sent to the winners on the publication date.

To enter, write down your answers to the following five questions:

1. What made you buy this collection of equipment reviews?

2. What was the last edition of the World Radio TV Handbook you purchased?

3. What is the brand and model number of your current receiver?
4. How would you judge the technical level of this first Buyers Guide? Too much technical detail, about right, or not enough graphs and test results?

5. Do you think we should include more accessory tests? If so, what kind of accessories do you have in mind?

Add your full name and address, and send the entry to Equipment Prize Survey, WRTH, P.O. Box 9027, 1006 AA Amsterdam, The Netherlands. The prize drawing will be made on August 15th 1993 and winners notified by post. All entries become the property of the WRTH and no correspondence will be entered into.

CREDITS

The Equipment Buyers Guide Edition One was compiled by Jonathan Marks and Willem Bos. Additional contributions by Richard Dixon, Lou Josephs, Dave Rosenthal, Mosche C Satt, Tom Sundstrom and Rocus de Joode. Additional thanks go to David Ward, Marian Meeuwissen, Fred Haanebeek, Danielle Hamstra, Hos van Hardeveld of Radio Communications Centre, Utrecht and Dr Kim Andrew Elliott for their background research. Also thanks to the more than 4000 WRTH readers who returned the 1992 Reader Questionnaire. Feedback, as always, is highly appreciated.